Recycling and Lifetime Management in the Textile and Fashion Sector

Editor

Kirsi Niinimäki

CRC Press
Taylor & Francis Group
Boca Raton London New York

CRC Press is an imprint of the
Taylor & Francis Group, an **informa** business

Cover credit: Designs by Elina Onkinen and Kasia Gorniak are made from re-purposed workwear. Photo – Diana Luganski/Aalto University.

First edition published 2024
by CRC Press
2385 NW Executive Center Drive, Suite 320, Boca Raton FL 33431

and by CRC Press
4 Park Square, Milton Park, Abingdon, Oxon, OX14 4RN

Library of Congress Cataloging-in-Publication Data (applied for)

ISBN: 978-0-367-49083-6 (hbk)
ISBN: 978-1-003-04441-3 (ebk)

DOI: 10.1201/9781003044413

Typeset in Palatino
by Innovative Processors

Prewords: Recycling and Lifetime Management in the Textile and Fashion Sector

The linear way of producing (make-use-waste) textiles and garments has expedited material throughput in the system, increased the environmental impacts of the industry, enlarged our wardrobe sizes and filled landfills with textile waste. The fast fashion business model has led to garment lifetimes being half of what they were 15 years ago and consumers buying 60% more garments than they did 15 years ago (Remy et al., 2016). The reason for this development has been the opening up of global business, effective and cheap mass-production in lower cost countries (the global South), and effective marketing. Fast fashion has been a successful business model, but has had dramatic impacts on the environment (e.g. Niinimäki et al., 2020). In the linear system, resource usage is wasteful and unsustainable, thus we need to construct a new understanding of the system in the context of sustainability.

As more information on the environmental consequences of this current way of designing, manufacturing, doing business and consuming textiles and fashion is emerging, knowledge of sustainability is also expanding to include the lifetime management approach as well as the aspect of recycling. Recycling is an important part of the circular economy strategy. The circular economy, CE, has recently emerged as the main focus of textile and fashion sustainability transformation. CE has been identified in the EU and its latest policy work will be an important foundation for sustainable development in both production and consumption. The European Commission's CE action plan, adopted by the EU in 2015, defines CE as: ...*an economy [that] aims to maintain the value of products, materials and resources for as long as possible by returning them into the product cycle at the end of their use, while minimising the generation of waste* (European Commission, 2015; 2019).

CE does not only mean closing the material loop and recycling waste back into industrial processes as raw materials; it also includes approaches to extending product lifetimes through reusing, reselling, updating and repairing products. The aim of the latest textile strategy of the EU is to change the sustainable paradigm in textiles and fashion through extending product lifetimes, designing repairable and recyclable products, and pushing the circular economy onwards, giving guidance on how to include recycled materials in new product design (European Commission, 2022).

The aim of CE is to retain the value of the products, materials and resources circulating in the economy for as long as possible (EMF, 2012; 2015). CE also requires new systems thinking and wide-reaching collaboration among different actors. Swain and Sweet (2021, p. 4) highlight the following approaches while transitioning towards CE: *(a) manufacturers need to implement Circular Manufacturing Systems (CMS), which involves designing products that should last significantly longer and be designed for easy repair, refurbishing and remanufacturing; (b) consumers and households need to adapt to different consumption patterns and business models, such as sharing, renting or service models; and (c) at a societal level, policies and infrastructure for waste management that support pricing, taxation and fee models, education and communication etc., need to be implemented.*

Because of the transition towards CE, product lifetimes are receiving the attention of academic research, and many stakeholders and industries are more than eager to test the practical implications of the latest research knowledge. The transition towards new sustainability needs time, new knowledge and guidance towards suitable stepping stones to proceed on this journey. Many initiatives are already underway that combine academic research and industrial-level experimentation to study, test and implement new sustainable ideas, technological innovation, models and tools, manufacturing and design methods in industrial or business practices (e.g. the FINIX project to transform the textile industry in Finland). Moreover, scaling up projects that test and apply research findings from the laboratory to the industrial scale are emerging (e.g. the New Cotton project scaling up new recycling technologies, the T-REX project for recycling blends and engaging consumers in the transition to sustainability). Many of these developments are linked to textile waste recycling or product lifetime management.

Many projects and institutes have drawn up roadmaps for future sustainable textiles and the fashion sector. They have identified collaboration as one of the critical attributes for achieving success. For example, in the UK, WRAP (The Waste and Resources Action Programme) has been doing important work for many years, not only to disseminate the latest information but also to build a future-oriented understanding of the change needed for sustainable textiles and clothing. Their report, *Textiles 2030 Roadmap,* aims to help industry and society make *rapid, science-based progress in climate action and move to a more circular system* (WRAP, 25 April 2021). They highlight the following three levels in this development work:

1. Designing products to look good for longer and to be recyclable
2. Testing reuse business models such as re-commerce, rental, and subscriptions
3. Setting up partnerships to supply and use recycled fibres in new products

This book provides a wide landscape for recycling and product lifetime management in the textile and fashion sector and therefore touches on many issues, all of which are connected to founding a circular economy. Through this approach, the book attempts to deepen the understanding of the sustainability transformation, which is currently desperately needed but also already happening in the design, manufacture, consumption and recycling of textiles and garments. The textile and garment recycling and redefining product lifetimes of this transformation are at the core of the new understanding of sustainability.

This book is divided into four themes which cover: the design aspects, the use phase, recycling and the meaning of data and technology in CE. All these aspects are essential for enhancing the system level transformation. I wish to thank all the writers for their contribution to this book and I hope you as a reader will enjoy its content.

13th of January 2023, in Tuusula, Finland **Kirsi Niinimäki**

References

EMF (2012). *Towards the Circular Economy. Vol. 1: An Economic and Business Rationale for an Accelerated Transition*. Cowes, UK: Ellen MacArthur Foundation.

EMF (2015). *Delivering the Circular Economy: A Toolkit for Policy Makers*. Cowes, UK: Ellen MacArthur Foundation.

European Commission (2015). *An EU Action Plan for the Circular Economy*. European Commission. Retrieved from https://doi.org/10.1017 CBO9781107415324.004

European Commission (2019). *Report on the Implementation of the Circular Economy Action Plan*. European Commission. Retrieved from https://doi.org/10.1259/arr.1905.0091

European Commission (2022). *EU Strategy for Sustainable and Circular Textiles*. Brussels, 30.3.2022 COM(2022) 141 final, (online) https://eur-lex.europa.eu/resource.html?uri=cellar:9d2e47d1-b0f3-11ec-83e1-01aa75ed71a1.0001.02/DOC_1&format=PDF

Feather, B.L. (1984). *Men's Wear: Garment Fit, Quality and Care*. Home Economics Guide. University of Missouri-Columbia. https://mospace.umsystem.edu/xmlui/handle/10355/71450

FINIX Sustainable textile systems: Co-creating resource-wise business for Finland in global textile networks -project https://finix.aalto.fi/

New Cotton-project https://newcottonproject.eu/

Niinimäki, K., Peters, G., Dahlbo, H., Perry, P., Rissanen, T. and Gwilt, A. (2020). The environmental price of fast fashion. *Nature Reviews, Earth and Environment*, 1: 189-200.

Remy, N., Speelman, E. and Swartz, S. (2016). *Style that's Sustainable: A New Fast Fashion Formula*. McKinsey and Company, 1-6.

Swain, R.B. and Sweethy, S. (eds.) (2021). *Sustainable Consumption and Production. Volume II: Circular Economy and Beyond*. Swam, Switzerland: Palgrave McMillan.

T-REX Textile recycling excellence-project https://trexproject.eu/

WRAP (25 April 2021). *Textiles 2030 Roadmap*. https://wrap.org.uk/resources/guide/textiles-2030-roadmap

Contents

Part IV: Data and Technology

Textiles and Garment Lifetimes

Kirsi Niinimäki

Aalto University, Finland
e-mail: kirsi.niinimaki@aalto.fi

Introduction

Product lifetime is an emerging research field and a term that as yet is not well defined. We are used to talking more about a product's lifecycle, and lifecycle analysis (LCA) is a well-known tool for calculating a product's environmental impacts. It takes into account only certain limited stages, for example, manufacturing processes, until the factory gate, and excludes the use and disposal stages. The seminal publication *Longer Lasting Products: Alternatives to the Throwaway Society* edited by Tim Cooper (2010) has played a profound role in establishing and expanding product lifetime understanding. Yet the definition of product lifetime is still emerging and can be approached from many different viewpoints such as design and manufacturing, use, maintenance or disposal behaviour, repairing culture, waste recycling and new business understanding.

Product lifetimes can be studied from different viewpoints and based on different theories, but the real, deep meaning behind the system of designing-manufacturing-consuming-disposing and how this system influences product lifetimes requires holistic understanding. It might be easier to approach garment lifetime through a narrower view, for example, through the durability issue alone, but we know that a product's high quality and technical durability do not guarantee a longer lifetime. At the end of the day, consumers are the stakeholders who decide the product disposal stage. Therefore we try to approach the definition of garment lifetime from several viewpoints in this chapter and in this book. As Cramer puts it (Cramer, 2019, p. 74) 'the lens of the garment lifetime further draws attention to the shared responsibility of producers and consumers.'

The culture of consumption needs to change, the system understanding needs to change (do we make the extension of product lifetimes culturally, technically or economically possible or not?). Thus, bigger changes are required to slow down the textile and fashion system and to create a better balance between manufacturing and consuming.

The newest information in this context is provided by the latest doctoral dissertations. These also show the way to widening and deepening the discussion in the fashion field. In her dissertation, Marium Durrani (2019) claimed that garment mending practices do not only provide opportunities to extend a garment's lifetime but also offer consumers opportunities to learn about materials and quality issues. Furthermore, mending practices affect consumers' creativity and provide an opportunity to develop more sustainable taste in garments and even fashion. On the other side of knowledge-building (the design side), for example, Jo Cramer in her dissertation 'The Living Wardrobe: Fashion Design for an Extended Garment Lifetime' (2019) offers ways to focus on how to extend garments' lifetimes through design solutions. She emphasises the aspect of shared responsibility for *sustain-ability*, which is an active stance. She highlights that 'both designer and wearer are positioned relationally within the garment lifetime, which redefines the idea of ownership being central to the garment life cycle to a model that views the worn experience as one of temporary custody' (Cramer 2019, p. 225). This viewpoint gives a great deal of weight to the use experience, sensorial knowledge and consumer satisfaction issues in postponing garment disposal, which is also discussed by Mila Burcikova in her dissertation 'Mundane Fashion: Women, Clothes and Emotional Durability' (2019) (more in Chapter 4).

Having a wide view of textile and garment lifetimes is important, and focusing on technical durability aspects is not the only challenging task. We need to understand the social-cultural trends behind the sustainable-consumption-production paradigm (SCP), the social pressure from the current fashion system and the economic paradigm that builds on a certain type of product lifetime. In the context of sustainability, especially in a circular economy context, extending textile and garment lifetimes can also include aspects of reuse and recycling (reusing as a product or recycling as a material) which are important issues to include in our discussion on product lifetimes. This chapter will start with economic reality; the change from a linear to a circular system. After this, we discuss longevity, quality, the several lifetimes aspect and garment disposal.

Garment lifetimes in a linear system

The traditional industrial way of functioning is to extract raw materials and process them in a suitable form for industrial scale production, manufacturing products, use and disposal. This is the way that the linear economy works. The highest value of the product is at the point of purchasing, after which the value of the product or material decreases rapidly. The *take-make-dispose* model has been made hugely effective in large-scale industrial manufacturing in the Global South. Outsourcing manufacturing to lower-cost countries in the Far-East and Asia has also meant outsourcing the responsibility for the environmental impacts to the manufacturing countries. The fast-fashion model, 'is based on fast-changing trends that have caused very short use times of garments and increased waste problems in all Western countries' (Niinimäki and Durrani, 2020).

It is important to recognise that the linear-economy model is based on the planned obsolescence of products concept and understanding products as disposables and fast replacements (Burns, 2010; Niinimäki, 2011). In the linear economy, economic profit is tied to one-time selling alone, so products are not designed for long-term use. Recent development in the fashion sector has also shown that repairing products is no longer even worthwhile and that it is actually cheaper to throw away a garment instead of using repair services and to buy a new one instead. 'Accordingly most products are throwaway articles in the Western world. Increasing numbers of products do not last the optimum use time and are discarded prematurely' (Mont 2008; cited in Niinimäki 2011, p. 216). This means that we are throwing not only waste into landfills but also garments that still have a functional life left. This is the consequence of our Western way of living and consuming and the need for emotional experiences through purchasing new things. 'Through the new purchase event the consumer can again experience excitement, enjoyment, joy and pleasure, at least for a moment' (Niinimäki, 2011, p. 36). Fashion is based on fast-changing trends and constant aesthetic changes in our desires, which leads to constant changes in the content of our wardrobes. The underlying reason for this is the economic paradigm which relies on psychological obsolescence and consumers' ever-changing desires. According to Burns (2010), psychological obsolescence can be divided into aesthetic and social obsolescence. Consumer products that relate 'to our appearance and self construction are evaluated on both aesthetic and social grounds. Products have a symbolic meaning to us, connected to psychological satisfaction through an emotional response: e.g. clothing and fashion enable consumers to gain social acceptance, affiliation with particular groups, and emotional beauty experiences' (Niinimäki, 2011, p. 217). The symbolic side of fashion is emphasised in the liner economy through rapidly changing trends to increase sales and speed up fashion consumption. In the linear economy, temporality is the underlying attitude, and everything happens quickly: designing, manufacturing as well as consumption. Material throughput in the system is very fast; garment use times are short, and garments end up in landfills after a few months of active use time.

Design time is also short in industrial design work, as is the time to find and consider different options in, for example, material choices (e.g. Karell and Niinimäki, 2020). Furthermore, designers do not think deeply about the lifecycle aspects in the linear economy and do not consider re-use or recycling alternatives (Kozlowski, 2012, pp. 27–28). Furthermore, as the use times of garments are short, they do not need to be designed or manufactured in a way that enables long-lasting or durable use, and therefore, most of our garments, at least in the fast fashion sector, are designed to be laundered only 10 times (McAfee et al., 2004). This sets the current quality level of clothing quite low, and it has even been stated that consumers no longer even look for the best quality. So the quality level of garments is only average or even lower. This is enough, as consumers are looking more for the best price than high quality aspects when purchasing (Jackson and Shaw, 2009, p. 125). It seems that in the linear economy, garment lifetime is controlled through temporality aspects (emphasising fast and short) and emotional fashion needs (emphasising constant change). Of course, there are exceptions to this, and many companies are working hard to improve their product quality or are trying to keep it at a stable level while suppliers or materials change (e.g. Piippo et al., 2022).

The garment lifetime in a linear economy is designed to be short, and this has environmental impacts. It has been evaluated that one of the most critical aspects of trying to make a sustainable transformation is the garment's use phase and its length. The average lifetime of a garment in the UK is evaluated to be 2.2 years, but during this time, the active use time is only 1.5 months (Waste and Resources Action Programme, WRAP, 2012). Table 1 shows how even a slight extension to active use time creates an environmental improvement, and therefore, designing more durable garments and longer product lifetimes, as well as changing consumer behaviour can have a substantial effect on the current unsustainable fashion system.

Table 1. Extending the garment use phase enables a reduction in the environmental impacts of this sector (WRAP, 2012)

Extending use time	Carbon savings	Water savings	Waste savings
10% = 3 months	8%	10%	9%
33% = 9 months	27%	33%	22%

Product lifetimes in a circular system

In a system based on a linear economy, design and manufacturing practices and the fast fashion business model keep the use of resources and consumption rates high. This can be changed by a circular economy approach. This approach is very often connected to closing the material loop and handling waste issues only, but it also means slowing down and decreasing total manufacturing and consumption. Thus, it can be seen as a real option for sustainable transformation (e.g. Niinimäki, 2018). A circular economy does not mean producing and consuming the way we currently do; we have to challenge the whole system and build a new understanding of sustainability. It is believed that industry-led efforts in a circular economy concentrate on supply chain efficiencies and post-consumer textile reclamation, and approaches that slightly improve the existing linear system without challenging the scale of either production or consumption (Cramer, 2019, p. 254). Even more critical discussions will be needed in the future, as well as consideration of not only the efficiency aspect but also the sufficiency aspect; what will the fashion consumption level be if we take climate issues seriously (see, e.g. Coscieme, 2022)?

If we place the focus on use and how to extend it, our way of designing, manufacturing and doing business will drastically change and this will also have a significant effect on the environment. According to Stahel (2017) '[c]ircular thinking (extending and intensifying use, reuse, redesign, recovering waste) uses fewer resources, and preserves the water and environment, ending up in low-carbon use, less use of energy and fewer virgin materials, making circularity, therefore, more ecological than industrial processes in the linear model' (cited by Niinimäki, 2018, p. 37).

'In a circular economy, the value of products and materials is maintained for as long as possible. Waste and resource use are minimised, and when a product reaches the end of its life, it is used again to create further value. This can bring major economic benefits, contributing to innovation, growth, and job creation' (Kirchherr et al., 2018). The circular economy can be an opportunity for a new kind of value creation. Many Western countries are showing interest in the circular economy and the discussion is rising to the policy level. It is possible to create an economic system which is 'regenerative' and 'restorative' (Jones and Comfort, 2017), and the discussion can even turn to not only efficiency aspects in the system but also a new understanding of sufficiency (what is enough?). The transformation into a circular economy needs technological advancement (e.g. in textile waste recycling), new design and remanufacturing strategies (extending the use and using waste as a resource) and new business thinking.

As we know, it is possible to increase product quality and manufacture durable products, but currently there are obstacles to this which could be economic, systemic, institutional or psychological in nature (Mont, 2008). These issues, as well as consumer behaviour, need further research to find ways to overcome these barriers and extend garment lifetimes and close the material loop through reverse logistics and returning textile waste to industrial processes as a raw material. The circular economy approach will change the understanding of product lifetimes, but will give more value to materials, products, an extended use phase and material recycling. Textiles and garments can still have even monetary value at the end of their lives, as raw materials going back to industrial manufacturing processes in a circular economy context.

Recently, we have experienced the emergence of different kinds of trials to extend garment lifetime such as reuse, repair, upcycling, re-design, second-hand fashion and fashion libraries, which are all good ways of using garments for a longer time. All these need new business thinking, and furthermore industry needs to understand waste as part of their responsibility and include this aspect in their business model (e.g. offering repair services, providing second-hand fashion sales, organising reverse logistics).

Longevity

To better understand garment longevity we start the discussion with the reasons behind garment disposal. It is good to research consumer behaviour and the aspects that influence disposal: emotional reasons for garment disposal but also garment failure. Cooper et al. (2013) studied the general reasons for garment disposal and found that '[s]ome are discarded because the fabric appears worn (typically around the cuffs, sleeves, collar or knees), have been torn, stained or damaged in some other way, or have shrunk or have become faded over time. Components such as zips or elastic may have failed or buttons been lost' (Cooper et al., 2013, p. 14).

Laitala et al. (2015) also studied reasons for garment disposal and found that the main one was changes in the garment. These could be holes or tears, garments beginning to look old, stains, garments being worn out, faded colour, lost elasticity, dimensional changes or pilling. The second most common reason for garment disposal was changes in size and fit issues linked to the users' own bodies. Interestingly, 22% of women reported changes in their own body size but only 10% of men informed this as the reason for garment disposal. The third most common reason was change in tastes: developing a dislike of the current style, colour, pattern or print, and so on. The fourth category was situational reasons like having several similar or better garments, having no occasion to use a garment, and change in life situation (Laitala et al., 2015). These reasons are garment quality issues as well as emotional reasons behind garment disposal on the consumer side.

Obviously, quality considerations mean that durability should be at the centre of the discussion. Stahel describes how durability can currently be understood as an 'undesirable quality, while it represents an obstacle to replacement sales' (Stahel, 2010, p. 157). With this kind of mindset, politicians very often begin to speak about increasing consumption as society is facing economic crises in a time of recession. In this kind of paradigm '[p]ossible longevity is aborted' (Stahel, 2010, p. 157). Once again we return to the idea that durability and longer product lifetimes are tied to systems thinking and based on predominantly economic systems. Stahel offers an alternative to the current economic thinking (based on the industrial economy) in a 'functional service economy', in which the supply side 'exploits the existing stock of goods in order to make more money with less resources input' (Stahel, 2010, p. 157). Here, the sharing economy offers models, and the platform economy offers channels for producers to find ways to offer consumers products through renting, leasing, sharing and swapping. When product ownership stays on the producer side, the motivation to invest in durable and long-life products is strong, as is the economic incentive to prevent waste and the loss of resources.

Stahel points out that the focus should be on the 'service life' of products. To make this as long as possible, the attention in design should be on durability, function and performance (these will be discussed in more detail in Chapter 2). If an extended service life is the focus, we have to create a new system around it by designing repair services, perhaps even updating services and investing in second-hand sales. This creates an economic power shift from globalised manufacturing to more local actors (Stahel, 2010) and offers opportunities to build product-service-systems (PSS) instead of only one time sales. Several examples already exist of fashion companies who offer repair services or quality guarantees for their garments (e.g. Patagonia, Frenn).

Functional vs emotional durability

If the functional aspect is at the core of fashion (e.g. sports clothing, outdoor and camping clothing) it is easier to provide the function as a service and offer these kind of garments through leasing services. Yet, other kinds of fashion consumption have been growing in the Global North and consumers are accepting alternative ways of consuming fashion through fashion libraries (leasing) (e.g. Niinimäki, 2021). This is an opportunity to move towards ownerless fashion consumption, and still enjoy the change and the symbolic values of fashion (e.g. designer value, brand value): having several users offers the opportunity to extend the garments' lifetimes.

Emotional durability is another opportunity to extend the lifetime of a garment. Building a caring attitude towards the garments we own means taking good care of them and beginning to understand them (materials, style, cut, use comfort). Over time and active use, we begin to treasure them and even love these garments which we find impossible to throw away even if their functional time is over (e.g. Niinimäki and Armstrong,

2013). Yet few garments achieve this special role in our wardrobe and most of our garments are meaningful to us in some way only for short periods of time and it is easy to dispose of them, end the emotional bond and replace them with new items.

Both functional durability and emotional durability are important aspects to take into account when aiming for garment longevity. Here, highlighting the importance of extended product lifetimes can really help the transition to a circular economy.

Quality

Although extending garment lifetimes is the focus in sustainable development, and although we know that garment failure in most cases leads to garment disposal, it is worth touching on the issue of quality in this chapter. As mentioned earlier, in the current system, garment quality can only be average or lower (Jackson and Shaw, 2009). We know that it is technically possible to design high quality, but economically this might not be possible (as high quality increases the end price of the product). Quality can be defined in different ways. One approach was presented by Feather (1984) through the basic formula: Garment cost = style + workmanship + serviceability (use time). He further claimed that 'traditional men's clothes have been better made than women's because patterns change less radically and less often' leading to longer use time, meaning longer service time (Feather, 1984).

Piippo et al. (2022) studied the extended understanding of quality in a circular economy context. The aim of their study was to investigate how the definition of quality changes in the transition from a linear business model to a circular one. 'Timeless design, technically durable materials, high-quality assembly work, and product functionality are important factors for product quality. They are part of companies' quality assurance work, which creates a foundation for the products' extended use-time and a base for products to circulate in the system for longer' (Piippo et al., 2022, p. 12).

The quality process starts during the strategy phase, when a company defines the quality levels, goal(s), and actions to reach these goals. Moreover, these quality goals 'are defined and engineered into the product' (Piippo et al., 2022, p. 14). After the strategy work, during which the quality standard is set, design work starts. This considers all the aspects of the product, its manufacture and use, checking and testing during the manufacturing stages, logistics practices, and service and business activities, which communicate the responsibility of the company to partners (also suppliers) and consumers. Quality also includes the use phase. Good product maintenance practices are important for extending the use phase. Garment lifetimes can be extended through services (e.g.

repair, redesigning, updating services). The circular economy extends quality work to include the recycling and recovery phases and actions related to these (Piippo et al., 2022). This is in line with the aspect of extended producer responsibility (EPR), which is now under development in the European Commission. EPR means that producers and importers also have responsibility for the product after its useful lifetime is over, meaning responsibility for the product as waste.

This kind of extended understanding of quality work in a circular economy context can also provide a framework for understanding product lifetime and how to extend it. 'In addition, an ecosystem must exist that can take care of different recovery methods, after the initial use phase, such as product take-back, repair, sorting, and recycling. Therefore, the stakeholders of the ecosystem have to be known already in the strategy phase in order to guarantee successful end-of-life and recovery processes' (Piippo et al., 2022, p. 14). This argument shows that it is not possible to extend product lifetime without an existing and functioning infrastructure and system to handle textile waste.

It is also important to understand that recycling can extend a product's lifetime through material recovery. This is a new dimension in quality work in a circular economy, but it also deepens the understanding of product lifetime. When a company sets a predefined lifetime expectancy for their products it also determines the quality level connected to this lifetime (Piippo et al., 2022). In some companies' strategy work, product lines can even have different quality and lifetime expectations (Saramäki, 2021).

Piippo et al. (2022) also showed that quality work and lifetime planning are important parts of a company's strategy work. 'The decisions made at the early stages of product development have far-reaching effects that influence even end-of-product-life processes. In the later sales, use and recovery phases, the companies can support quality and lifetime extension through improved communication about quality and different care and recovery services. In this way, companies can even have an impact in these later phases in a garment's lifetime, which are traditionally thought to be out of their reach as the connection with the user is often lost after the sales transaction' (Piippo et al., 2022, p. 12).

The image of high quality garments also creates the attitude among consumers that they should take better care of these garments (e.g. McNeill et al., 2020; Saramäki, 2021). Good quality also enables extended use by having several users, and even enables second-hand sales. 'There is a likelihood that perceived quality positively influences attitudes toward garment life extension strategies' (Aakko and Niinimäki, 2021). Some companies (e.g. Emmy) provide a platform for consumers to sell their old garments to other consumers, trusting only brand garments to ensure high quality. They might still inspect these garments and even offer a quality category for second-hand items (e.g. Emmy) to build trust among the consumers who are buying these second-hand fashion items. We can argue that high quality is essential for lifetime understanding, lifetime extension and also the several lifetimes approach.

Several lifetimes

Extending the life of a garment has been highlighted as the most effective way of reducing the environmental impact of fashion (e.g. Sandin et al., 2019). This means that we have to find several alternatives for our existing throwaway culture in fashion. Different ways of extending lifetimes are needed to slow down material throughput in the system. Second-hand fashion is one of these alternatives, and needs new business understanding as well as a change in consumer behaviour. Second-hand fashion has gained popularity, especially during the COVID-19 crisis, and consumers seem to have become interested in new ways of consuming fashion (e.g. Niinimäki, 2021; Iran et al., 2022). Even if consumers' environmental interest is the main motivation behind the increase of second-hand fashion, Han and Sweet (2021, p. 174) claim that second-hand shopping is largely driven by the opportunity to gain symbolic and cultural capital. Fashion is about constant change through ever changing trends, but it is also about symbolic capital through brands and owning. But in second-hand fashion, values like individuality, authenticity, and distinction can also be emphasised (Han and Sweet, 2021). This highlights the importance of creating a new fashion culture in a circular economy. A new understanding of garments' extended lifetimes and garments having several lifetimes needs to be constructed. A culture of good maintenance and repair, appreciating the material world again, and even ownerless consumption by, for example, using fashion libraries to bring about change and fashion experiences, needs to be cultivated for a sustainable transition towards longer garment lifetimes.

The several lifetimes approach, i.e. a garment having several users, can be implemented by new kinds of business models – re-selling, re-using, updating and remanufacturing – but also by lending and leasing. It is obvious that the several lifetimes approach needs garments of high quality so that they retain as high a value as possible for as long as possible. For example, leasing and renting keeps the ownership of garments on the business side, and garments are investments that need to be used for as long as possible to make the business profitable.

Optimal lifetime

Saramäki (2021, p. 22) listed the ideal lifetime of different product groups on the basis of the recommendations of The Federation of Finnish Textile Services (Tekstiilihuoltoliitto ry) as follows: 'Classic skirt suits, classic formalwear, coats and outdoors outfits 5 or 6 years; men's suits, skirts, men's blazers and lightweight poplin coats are expected to last

4 years; trousers, trouser suits, blouses, shirts, knitwear, trendy skirt suits, trendy formalwear and dresses nightwear 3 years; and trousers, jeans, sportswear, home and free time wear for 2 years' (see Table 2). These 'optimal lifetimes' are actually the minimum time for which we can expect a garment to last. This list is based on the expectation that these times concern 'quality garments' and therefore exclude fast fashion garments. In its sustainability agenda, WRAP in turn has given guidance on garments' long lifetimes by stating that the average lifetime is 4.5 years (in more detail, 113 wears and 56 laundry washes) (Cooper et al., 2014).

Table 2. Ideal lifetimes of different garment types (according to Tekstiilihuoltoliitto ry)

Ideal lifetime	Product category
5–6 years	Classic skirt suits, classic formalwear, coats and outdoors outfits
4 years	Men's suits, skirts, men's blazers and lightweight poplin coats
3 years	Trousers, trouser suits, blouses, shirts, knitwear, trendy skirt suits, trendy formalwear and dresses nightwear
2 years	Trousers, jeans, sportswear, home and free time wear

We all have a favourite garment that we use all the time while others remain unused in our wardrobes. Niinimäki and Armstrong (2013) noticed that emotional bonding might extend the time for which the garment is owned, but not its active use time (see Table 3). The elements that create attachment to garments are functionality, memory dimensions, emotional satisfaction, design and style, fabric and material, personal values, quality, effort in investment and financial value. The study also found that wearing frequency affected the lifetime of clothing. Six years of owning garments meant active use either daily, weekly or several times a month. Jeans, active wear, T-shirts and dresses were most actively used. Interestingly, there was a turning point at which active use and use satisfaction changed to memories. 'The results from the questionnaire show that in garments owned between 1–3 years, a few respondents (17 respondents of 99) reported some memory connection while most other responses showed that meaningful attributes were linked to functionality, beauty, and comfort in use. Garments in frequent use also elicited aesthetical experiences and these were linked to the wearer's self-image and emotional satisfaction' (Niinimäki and Armstrong, 2013, p. 196.) This study found four stages of owning meaningful garments (see Table 4). When a garment is owned for 0–6 years, the focus is on use enjoyment (see Table 4). Liking and loving a garment begins when it has been owned for 7–16 years. After being

Table 3. Length of time garment is owned when owner is deeply emotionally attached to it (based on Niinimäki and Armstrong 2013, p. 195)

Product type	Length of time owned (years)		
	Minimum	Maximum	Mean
T-shirt/Sweat shirt	0.2	44	11.000
Dress	0.3	50	9.659
Jeans or casual pants	0.2	35	5.398
Active wear	0.5	30	5.735
Accessory	0.5	31	7.346
Sleepwear/Undergarment	3	23	10.077
Outwear	1.5	44	15.875

Table 4. Stages of attachment to garments (according to Niinimäki and Armstrong, 2013)

Stage	Use enjoyment	Liking and loving	Reflective relationship	Cemented memento
Years item is owned	0–6	7–18	19–21	22<

owned for 17–21 years comes the reflection stage; garments are seldom any longer in use, but have gained a great deal of emotional value. This study also showed how garments that are owned for over 22 years become 'cemented mementos' and are passively kept in storage and no longer actively used (Niinimäki and Armstrong, 2013).

Few garments achieve a long lifetime, and most are disposed off after a short time of being used or owned (average time owned, two to four years), but this study shows the importance of emotional bonding and its connection to long product lifetimes. On the other hand, good functionality and an enjoyable use experience in the active use phase may possibly extend use time and build emotional bonding with the garment. 'The meaningful clothing item had offered a pleasurable use experience and deep satisfaction to the wearer during the active wearing stage' (Niinimäki and Armstrong, 2013, p. 197). Emotional bonding needs time and reflection to emerge. As Russo (2010) argues, through reflection, emotions might arise and users may begin to become attached to products through feelings of love, passion, and commitment. We love some of our clothing deeply and wear it regularly, we have passion for fashion or at least we strive for aesthetic experiences through garments and fashion. When a garment has reached the attachment stage, it is cherished and we take good care of it (we feel commitment towards it) (Niinimäki, 2010). Yet we know that high quality, technical durability or even emotional bonding might not save the garment from early disposal. As Van Nes and Cramer (2005, p. 287) put it, 'product lifetime is a result of a user's decision, and not a predetermined design criterion.'

Goldsworthy (2017) talks about 'lifetime optimization', in which lifetime extension is not always the aim, but critical evaluation decides the optimal lifetime, when the environmental burden of the product or its use is taken into account. Normally this kind of discussion concerns white goods or other products that use electricity, and in this case newer products are more energy efficient than old ones. Could this concept somehow be applied to textiles or fashion? Perhaps in the use of polyester, it could. Polyester is a part of the microplastic problem in the oceans (Niinimäki et al., 2020) and we need more knowledge on how to avoid this problem. Should we use polyester garments for only a short time so that less microplastic would be released during laundry or should we totally ban the use of polyester because of its environmental impact? On the other hand, polyester is a strong fibre, which enables long use, and in many use contexts it is quite durable (like sports clothing). It would be beneficial to further study how to optimise garment lifetimes. Garments that are meant to be used for very long times should be made with the highest quality materials so that they can achieve the best service life. On the other hand, if we could better predict the optimal use time of a garment, we could select the optimal quality level for it and choose materials accordingly (thus saving resources and costs). These kinds of garments with a shorter lifetime could end up in effective material recycling at the end of their life and perhaps need minimum washing during their use phase (thus reducing the environmental impact during its use phase). Here it might be useful to include two different approaches: the lifetime of the product and the lifetime of the material. It would be beneficial to understand and define the length of the product lifetime and select materials according to, for example, the wash times this garment needs to be able to endure its use time without wearing and tearing. The lifetime of the material could also be calculated, but this should include the aspect of recycling and how many times the material can be recycled. Recycling is important in the circular economy context.

Conclusions

We need to focus more on product lifetimes and the environmental impact of a product. Product lifetime is not an easy issue; it is not possible to define in a simple way, as shown in this chapter. This topic needs more research, and based on this knowledge, different kinds of industrial strategies could be constructed. Another aspect is how producers could include textile/garment waste as part of their business model and in this way extend their responsibility throughout the garment lifetime. The end of a product's lifetime would then also be the responsibility of the producer/manufacturer. Investing in recycled textile material is a good way to reduce the environmental impact of manufacturing instead of using virgin materials only. Yet new research is still required to calculate the benefits of not using or using fewer virgin materials. Moreover, LCA is needed for different recycling methods.

Industrial tempos need to change, from fast to slow and from short to long. The circular economy might provide us with a new understanding of this and help us focus on longer use times but also on the recyclability of products and materials. Recycling extends product lifetimes even after the use phase is over and the product's life continues in a material form. This is a new stage that should be included in our sustainable design understanding. In the future, we will have to design product lifetimes instead of just products. Re-use, repair, re-design, up-cycling are all important strategies that could much more confidently be used to extend product and material lifetimes as part of current textile and fashion business.

Acknowledgements

This research was supported by the Strategic Research Council at the Research Council of Finland, Grant no. 352616 FINIX consortium.

References

Aakko, M. and Niinimäki, K. (2021). Quality matters: Reviewing the connections between perceived quality and clothing use time. *Journal of Fashion Marketing and Management*, 26(1): 107-125.

Burcikova, M. (2019). *Mundane Fashion: Women, Clothes and Emotional Durability.* PhD thesis, School of Art, Design and Architecture, University of Huddersfield, UK https://ualresearchonline.arts.ac.uk/id/eprint/15829

Burns, B. (2010). Re-evaluating obsolescence and planning for it. In: T. Cooper (Ed.), *Longer Lasting Products: Alternatives to the Throwaway Society*. pp. 39–60. Farnham: Gower.

Cramer, J. (2019). *Living Wardrobe*. Doctoral dissertation, RMIT. https://researchrepository.rmit.edu.au/discovery/delivery/61RMIT_INST:ResearchRepository/9921863896801341#13248374120001341

Cooper, T. (2010). *Longer Lasting Products: Alternatives to the Throwaway Society*. Gower, Farnham: UK.

Cooper, T., Hill, H., Kinmouth, J. and Townsend, K. (2013). *Design for Longevity: Guidance on Increasing the Active Life of Clothing*. WRAP, UK. https://wrap.org.uk/resources/report/design-extending-clothing-life

Cooper, T., Claxton, S., Hill, H., Holbrook, K., Hughes, M., Knox, A. and Oxborrow, L. (2014). *Clothing Longevity Protocol.* Nottingham Trent University, Banbury/ WRAP.

Coscieme, L., Akenji, L., Latva-Hakuni, E., Vladimirova, K., Niinimäki, K., Henninger, C. et al. (2022). *Unfit, Unfair, Unfashionable: Resizing Fashion for a Fair Consumption Space.* Hot or Cool Institute, Berlin. https://hotorcool.org/ unfit-unfair-unfashionable/ .

Durrani, M. (2019). *Through the Threaded Needle: A Multi-sited Ethnography on the Sociomateriality of Garment Mending Practices.* Doctoral dissertation, Aalto University, Finland https://aaltodoc.aalto.fi/handle/123456789/41489

Feather, B.L. (1984). Men's wear: Garment fit, quality and care. *Home Economics Guide.* University of Missouri – Columbia. https://mospace.umsystem.edu/ xmlui/handle/10355/71450, accessed 10.7.2021

Goldsworthy, K. (2017). The speedcycle: A design-led framework for fast and slow circular fashion lifecycles. *The Design Journal*, 20(sup 1): S1960-S1970, DOI:10. 1080/14606925.2017.1352714

Han and Sweet (2021). Consumers practicing sustainable consumption: Value construction in second-hand fashion markets. *In:* Swain, R.B. and Sweethy, S. (Eds.), *Sustainable Consumption and Production*, Volume II: Circular Economy and Beyond, pp. 180-193. Swam, Switzerland: Palgrave McMillan.

Iran, S., Joyner Martinez, C.M., Vladimirova, K., Wallaschkowski, S., Diddi, S., Henninger, C.E. et al. (2022). When mortality knocks: Pandemic-inspired attitude shifts towards sustainable clothing consumption in six countries. *International Journal of Sustainable Fashion and Textiles*, 1(1): 9-39.

Jackson, T. and Shaw, D. (2009). *Mastering Fashion Marketing.* New York: Palgrave Macmillan.

Jones, P. and Comfort, D. (2017). Towards the circular economy: A commentary on corporate approaches and challenges. *Journal of Public Affairs*, 17(4): 1680.

Karell, E. and Niinimäki, K. (2020). A mixed-method study of design practices and designers' roles in sustainable-minded clothing companies. *Sustainability*, 12(11): 4680.

Kirchherr, J., Piscicelli, R., Bour, R., Kostense-Smit, E., Muller, J., Huibrechtse-Truijens, A. and Hekkert, M. (2018). Barriers to the circular economy: Evidence from the European Union (EU). *Ecological Economics*, 150: 264-272.

Kozlowski, A., Bardecki, M. and Searcy, C. (2012). Environmental impacts in the fashion industry: A life-cycle and stakeholder framework. *Journal of Corporate Citizenship*, 45: 17-36.

Laitala, K.M., Boks, C. and Klepp, I.G. (2015). Making clothing last: A design approach for reducing the environmental impacts. *International Journal of Design*, 9(2): 93-107.

McAfee, A., Dessain, V. and Sjoeman, A. (2004). *Zara: IT for Fast Fashion.* Boston: Harvard Business School Publishing.

McNeill, L.S., Hamlin, R., McQueen, R.H., Degenstein, L., Wakes, S., Garrett, T.C. and Dunn, L. (2020). Waste not want not: Behavioural intentions toward garment life extension practices, the role of damage, brand and cost on textile disposal. *Journal of Cleaner Production*, 260: 121026.

Mont, O. (2008). Innovative approaches to optimising design and use of durable consumer goods. *International Journal Product Development*, 6 (3/4): 227-250.

Niinimäki, K. (2010). Forming sustainable attachment to clothes. *7th International Conference on D and E Conference* in IIT , 4-7 October 2010, Chicago, USA.

Niinimäki, K. (2011). From Disposable to Sustainable. The Complex Interplay between Design and Consumption of Textiles and Clothing. Doctoral dissertation. Helsinki: Aalto University. Ihttps://aaltodoc.aalto.fi/handle/123456789/13770, accessed 10.7.2021

Niinimäki, K. and Armstrong, C. (2013). From pleasure in use to preservation of meaningful memories: A closer look at the sustainability of clothing via longevity and attachment. *International Journal of Fashion Design, Technology and Education*, 6(3): 190-199.

Niinimäki, K. (Ed.) (2018). *Sustainable Fashion in a Circular Economy*. Aalto ARTS Books.https://aaltodoc.aalto.fi/handle/123456789/36608, accessed 10.1.2021

Niinimäki, K. and Durrani, M. (2020). Repairing fashion cultures: From disposable to repairable. *In:* L. McNeill (Ed.), *Transitioning to Responsible Consumption and Production*. pp. 154-168. Basel Switzerland: MDPI Publication. https://www.mdpi.com/books/pdfview/edition/1246

Niinimäki, K., Peters, G., Dahlbo, H., Perry, P., Rissanen, T. and Gwilt, A. (2020). The environmental price of fast fashion. *Nature Reviews; Earth and Environment*, 1: 189-200. https://doi.org/10.1038/s43017-020-0039-9, accessed 10.1.2022

Niinimäki, K. (2021). Clothes sharing in cities: The case of fashion leasing. *In:* J. Corcoran and T. Sigler (Eds.), *The Modern Guide to the Urban Sharing Economy*. pp. 254-266. Cheltenham Glos, UK: Edward Elgar Publisher.

Piippo, R., Niinimäki, K. and Aakko, M. (2022). Fit for the future: Garment quality and product lifetimes in a CE context. *Sustainability*, 14(2): 726.

Russo, B. (2010). *Shoes, Cars, and Other Love Stories. Investing the Experiences of Love for Products* (Dissertation). Delft University of Technology, Delft.

Sandin, G., Roos, S., Spak, B., Zamani, B. and Peters, G. (2019). *Environmental Assessment of Swedish Clothing Consumption* (A Mistra Future Fashion Report No. 2019:05). Retrieved from http://mistrafuturefashion.com/wp-content/uploads/2019/08/G.Sandin-Environmental-assessment-of-Swedish-clothingconsumption. MistraFutureFashionReport-2019.05.pdf.

Saramäki, R. (2021). *Communicating Clothing Quality*. Master thesis, Aalto University, Finland. https://aaltodoc.aalto.fi/handle/123456789/108402

Stahel, W. (2010). Durability, function and performance. *In:* Cooper, T. (Ed.), *Longer Lasting Products: Alternatives to the Throwaway Society*. pp. 157-177. Gower, Farnham: UK.

Stahel, W. (2017). Preface. *In:* Baker-Brown, D. (Ed.), *The Re-use Atlas*. pp. xiii–xviii. London: Riba.

Tekstiilihuoltoliitto ry (2017). Nykyarvon määräytymisperusteet, tekstiili todennäköinen käyttöikä laadukkaille tekstiileille", http://tekstiilihuolto. web31.neutech.fi/doc/TODENNKINEN-IK-NYKYARVONMRYTYMISPER USTEET-2017-13062017.pdf, accessed 10.1.2021

Van Nes, N. and Cramer, J. (2005). Influencing product lifetime through product design. *Business Strategy and the Environment*, 14: 286-299.

WRAP, Faye Gracey and David Moon (2012). Valuing our clothes: The evidence base. http://www.truevaluemetrics.org/DBpdfs/Issues/Supply Chain/Textiles/WRAP-REPORT-10.7.12%20VOC-%20FINAL.pdf, accessed 10.1.2021.

PART I
Design

Designing for (Extended) Product and Material Lifetimes

Kirsi Niinimäki

Aalto University, Finland
e-mail: kirsi.niinimaki@aalto.fi

Introduction

Traditional garment design decisions require information on aesthetics, cost, quality and function, as well as information on the business niche and psychological consumer needs (Kozlowski et al., 2012, p. 24). However, this approach needs to be expanded when the consideration of product lifetime steps into play. 'We need to find ways to understand the potential impacts of each and every design decision we make, and the specific attributes of the products we are designing in relation to their life-journey' (Goldsworthy, 2017, p. S1963). Lifetime and lifecycle thinking create new challenges for designers. As Kozlowski et al. (2012) highlight, a more strategic approach is important, in which 'incorporating environmental, social and stakeholder considerations…can lead to strategic supply chain management to support the company's environmental and social goals.' They also point out that this strategic approach can help us think about the use phase and how to support its longevity (how to take care of and maintain the product in the best way), reuse the product, and even its end-of-life management. Widening the perspective from designing a product to designing lifetimes changes the mindset and paradigm in design, and a company's strategy work can strongly support designers' work (Kozlowski et al., 2012, pp. 26–27).

Cooper et al. (2013, 14) claim that less resources (time and money) are used in the design and manufacturing stages if the item in focus is an industrially mass-manufactured garment, and that this is why the quality and durability of garments are uncertain. They also claim that 'some garments last for just a few washes'. The struggle for cost savings in the fashion business is hard, and therefore cheaper fabrics and processes are often chosen in industrial manufacturing. This approach might be frustrating for designers: 'their work is being treated as a throwaway item, yet for retailers under pressure to improve profits, the concept of longevity implies lower sales volumes and revenue, or higher prices' (Cooper et al., 2013, p. 14).

A garment's lifetime is a complex issue and has two main aspects (a) attributes that link to a product, and (b) consumer behaviour. This chapter mainly focuses on the product side, but also touches on issues linked to the use phase. It aims to present the design world and discuss the issues on the designer's desk; basically what issues the designer can work with, and which are out of the designer's control. The previous chapter already mentioned some issues linked to garment disposal (e.g. failure in technical durability), but we know that early disposal of garments is also quite a common practice: garments are not thrown away because of some technical failure but because consumers are tired of them or have emotional need for change.

Technical durability

Technical durability is one of the most important factors for evaluating quality and aiming to design a product with a long lifetime. Cooper and Claxton (2022) have studied garment disposal and the reasons behind it. They have categorised the failures of different garment types (see Table 1) and found that they are related to failures in technical quality or design. Designers can learn from these failures, and thus it is important that they collect customer feedback. Very often, a new garment will be tested by the designer himself or some other person in the company, implying that a new material or new product can be tested through real user experience before it gets to the market. However, this approach requires time in the product development phase.

It is hard to control all aspects and details while designing a new product. Even acquiring information to be able to compare alternatives is hard. It is even harder to estimate the longevity aspects of materials or sewing work while making design decisions. In most cases, designers are able to control their material choices, but they might not have enough time to find different material options or cost limitations may prevent the selection of the most suitable material. Technical tests provide some numerical information about the durability of the material (Karell and Niinimäki, 2020), and material durability can be evaluated through standardised testing (e.g., ISO9000). For example, colourfastness (wash resistance and lightfastness), dimensional stability, and abrasion resistance are often tested (Bubonia, 2014).

Table 1. Types of physical failure of garments discarded by their owners (Cooper and Claxton, 2022, p. 5)

Garment	Failure type
Cardigans	Pilling, faded colour, loss of dimensional stability
Shirts	Discolouration, faded colour, fabric breakdown
T-shirts	Pilling, faded colour, logo failure, loss of dimensional stability
Jeans	Accidental damage [stain, rip], (unintentional) faded colour, holes in seams
Formal trousers	Fabric breakdown, holes in seams
Jackets	Faded colour, holes in seams

A recent study (Karell and Niinimäki, 2020) has shown that even sustainable-minded companies and their designers have difficulties finding eco-material that is most suitable for their product type. Functionality and durability are at the core of fashion designers' material choices and in most cases, longevity is taken seriously at this stage. No compromises can be made by, for example, choosing an eco-material with lower durability, because durability is one important way to build trust among consumers in the use phase and this aspect is strongly linked to brand reputation. To find suitable materials and ensure the continuation of material delivery many companies actively participate in material development processes (i.e. long-term collaboration with a textile factory). Some companies have noticed variation in material quality from the same supplier. The choice of sufficiently high and stable material quality are critical aspects of product longevity (Karell and Niinimäki, 2020).

It is important that the basic design attributes of garments, such as pilling resistance, colourfastness and dimensional stability are controlled through design decisions so that long use time can at least be offered to users (Gwilt and Pal, 2017) or to enable extended use by several users (e.g. reuse, second-hand fashion, renting, leasing). This is especially important in the circular economy context, in which the longevity of a product is a value. Furthermore, material is worth repairing if it is of a higher quality, and it is easier to recycle at the end of the product's lifetime if the original material is of a higher quality.

Quality work in companies is very important from the technical durability viewpoint. Different quality requirements are set depending on the product type, for example, whether it is children's wear or knitwear. Different product categories have different durability expectations: some garments are washed frequently and others need high rubbing durability. Jo Cramer (2019, p. 73) explains that physical durability means developing and designing products that can 'take wear and tear without breaking'. Here, material selection during `the design phase is critical. Other details are also important, such as trimmings, haberdashery, fabric finishing, dyes. Cramer further points out that *design for reliability* includes the use phase, and highlights the importance of a product's reliability during the pre-planned use time, based on the idea that each product type has an ideal or optimal use time. During this time, the product should not fail, but consumers' knowledge and activity is needed to safeguard long use time. Consumers should follow maintenance instructions of the product properly for it to last the planned optimal lifetime (Cramer, 2019). If companies want to emphasise the high quality of their garments, they can offer warrantees.

When we talk about technical quality, most of the time we are referring to aesthetics, that is, wear and tear not showing. The materials look good and do not change much during their use time. Materials can also age 'gracefully', meaning that 'patina' – the wear – just adds to the beauty of the material. For example, real leather is considered a material that improves in appearance as it ages and the ageing process is visible (Niinimäki and Koskinen, 2011). The ageing process of jeans is accepted by consumers as part of the product image. In the case of jeans, even the fading of the colour (indigo colour) is part of the charm and totally accepted as part of the product's 'character'. This approach shows that by selecting suitable materials, designers can even include the aspect of ageing in the product's lifetime, and it could be one design element in their work. Currently, knowledge on the ageing process of different material is limited. As Lilley et al. (2019, p. 417) explain '[h]owever, materials resources for designers rarely provide information about how materials will change in use.' This could be new design knowledge in the future; either for designing garments that endure wear and tear with no visible changes or garments for which the change is part of the product's aesthetics and attractiveness.

Designing longer-lasting style

Other aspects that fashion designers consider when they want to create longevity are avoiding styles that are too trendy or creating a design approach that totally avoids seasonal trends. This might mean favouring timeless design (Karell and Niinimäki, 2020, 6) or slow fashion by slowly changing or evolving collections (Henninger et al., 2022).

'Multiple sizing, transformability or loose pattern design are some practical means' to achieve longevity (Karell and Niinimäki 2020, 6). Cooper et al. (2013, 16) also note that 'oversized knits and kimono shapes that can be worn with a belt were described as versatile and "comfortable", therefore potentially wearable for longer.'

Some companies and designers focus on classic items and classic style (like the little black dress, LBD) and classic colours (e.g. black) which enable longer sales times for these garments (see Table 2). These items do not go out of style. This also enables designers to slow down the design process and focus on, for example, increasing the quality of these classic pieces. When the production is based on more stable design, there is more time to concentrate on increasing quality and improving details (Cooper et al., 2013, p. 16).

Table 2. Examples of classic garments and basic colours

	Garment/Colour	Reference
Classic garments	Little black dress Tailored shirts Pencil skirt Chino-style trousers Jeans V-neck jumpers	Cooper et al., 2013
Classic garments	Suits Denim trousers and a jacket Trench coat White shirt Pencil skirt Little black dress	Verbič, 2019
Core colours of garments	Black White Navy Grey Red	Cooper et al., 2013
Neutral colours of garments	Blue Black Grey White	Durrani and Niinimäki, 2023

Adjustable sizing

One of the most common reasons for disposing of garments is changes in body size. As Laitala et al. (2015, p. 7) found, in consumers' garment disposal decisions '[t]he next largest main group was related to problems with size and fit, either that the owners had grown out of their clothing, or that the clothing never fitted well to start with.'

Body size fluctuation could be taken into account in design. Currently, most of our garments are made of blended materials that contain a certain amount of elastic, which in turn gives a certain stretch and provide not only use comfort but also flexibility in size. However, it is good to remember that material blends, especially elastane, are problematic from the recycling point of view (see Chapter 9), and thus adding elastic material might lengthen the garment's lifetime but prevent its recycling.

However, there are other design options to make the size of a garment adjustable. Providing extra material in the seams (wider seam allowance) enables changing the size of the garment if the wearer's body size changes or if another user of a different size uses the garment (Laitala, et al. 2015; Ikävalko, 2022). Other ways are focusing on the diagonal cut of the fabric, a wrapping design, a modular design, or designing garments with a loose fit (even one size only) and to be used in several ways, perhaps by building in some visible or invisible tightening system inside or outside the garment (Ikävalko, 2022). As a part of her Master's thesis, Ikävalko (2022) designed different simple ways to control size by including hooks, ribbons, hidden and resewn seams or pleats (Ikävalko, 2022) and created an aesthetic and functional way to extend garments' lifetimes. Moreover, Karell (2014) developed and experienced a modular fashion design, the idea of which was to enable modifications during the garment's lifetime. This study needed close consideration of what kind of textile materials can be used in this type of garment so that the construction can be dissembled and then sewn together again. Karell selected high-quality woollen crêpe, as this allows unstitching marks to be steamed away before the garment is re-assembled (2014).

The made-to-measure approach by which garments are made according to the consumer's own individual size could be combined with 3D body scanning to offer the perfect fit and thus consumer satisfaction; a service to consumers who consider the perfect fit to be the most important aspect of garments. Adjustable size could also be included in this design strategy.

Designing for use satisfaction

The fit of a garment is an important aspect of garment satisfaction. Tailored or semi-tailored and different strategies for made-to-measure garments are one way of offering the garment user better satisfaction. These approaches offer well-fitting garments, which can be understood to 'frame the form well aesthetically' (Cooper et al., 2013, p. 16). Garment quality is linked to its material elements, such as fabrics, construction, colour and finishing, but also to its design features, such as fit, functionality and performance (De Klerk and Lubbe, 2008) and all these quality elements also influence the wearer's use experience. As previously stated, the visual appearance of a garment is not the only aesthetic experience that is important for the consumer: another level is the use experience leading to satisfaction or dissatisfaction. Excellent drape, lustre, softness (Cooper et al., 2013, p. 15), the weight of the material on our body, or a pleasant kinetic and tactile interaction with the garment are part of the beauty experience of the apparel (Niinimäki, 2014). 'The wearer can experience aesthetic satisfaction with a suit or evening dress that fits him/her perfectly and is made of high-quality material, using a high level of technical skill' (Niinimäki, 2014, p. 3.5).

Some previous studies have also looked at consumer experiences of garments. 'Physical comfort in clothing continues to be a high priority for men. Whether in a business suit or casual jeans, the way clothes fit affects how the man projects himself. Researchers have found men rated comfort and fit to be the most important factors in wear studies and preference evaluations. The comfort ratings were found to be dependent on the fit of the garment. Even in the most casual clothing, fit is critical to appearance and comfort' (Feather, 1984, p. 712). Other studies of female consumers have also pointed out that not only is the visual experience important for women, but physical and bodily aspects are also part of the beauty of the garment. Female consumers also more often 'yearn for beauty' (54% of female, 6% of male consumers) when purchasing a garment. The fit aspect is important for both genders (65% for male, 67% female consumers) (Niinimäki, 2018, p. 11). Overall, (without dividing consumers) the experiences that garments offer us are an important part of use satisfaction for everyone. The study highlighted the following as the most important reasons for garment disposal: damage/quality problems, lack of use, wearing out, uncomfortable in use, needs too much maintenance (Niinimäki, 2015). Avoiding these failures through the right design decisions offer possibilities to extend longevity.

A study by Niinimäki and Armstrong (2013) show the importance of functionality for meaningful garments to be used for a long time. The consumers named the following attributes as important in forming an attachment to a garment; comfortable, good fit, warm, multi-functional, functional (good for sports and hiding body deformations), easy to match, easy to care for and easy to put on. They also described emotional satisfaction, design and quality issues as part of use satisfaction (see Table 3). These kinds of attributes may mean that a garment is worn more frequently but can also mean that the garment might be kept for a longer time, thus extending the garment's lifetime. This extension of owning and use time was evident in the study.

New design strategies and services to extend garment lifetime

Recover, recycle, repurpose, remanufacture, refurbish, repair, reuse, reduce, rethink, and refuse are ten relevant strategies in the circular economy context (Morseletto, 2020). All these approaches challenge the current way of designing, manufacturing and consuming garments. Re-making or re-designing means using fashion waste, discarded garments or waste textiles as a material for new garment design. 'Up-cycling tends to describe a process that re-makes the garment into an entirely new

Table 3. Elements and attributes that designers could work on to extend garment lifetime (based on Niinimäki and Armstrong, 2013, p. 216)

Category	Attribute
Functionality	Comfortable Good fit Warm Multi-functional Functional (good for sports and hiding body deformations) Easy to match Easy to care for Easy to put on
Emotional satisfaction	Look good in it Feel good in it Get compliments when wearing it Best piece Love the brand
Design	Good design (beautiful, pretty, and cool-looking) Stylish Classical and timeless design
Material and colour	Nice colour Tactile feeling (silky and soft) Fabric aesthetic (not thick, light weight, and sparky) Flexible (not stretched)
Quality	Durable High quality manufacturing High quality material

form that may or may not exhibit traces of the original garment' (Cramer, 2019, p. 243). Cramer (2019) uses the term re-modelling for all different kinds of approaches that involve making alterations to a garment to maintain its current function, and the term re-design for deconstructing and reassembling the garment into a totally new form. Up-cycling also means an increase in value, for example, designing new products with monetary value from valueless waste material.

All these approaches are important if the aim is to extend the use time of a garment or to facilitate the recycling of textile materials (extending their use as valuable materials in the industrial system). These strategies could be used as design tools as well and may also create services related to them. The service aspect enables extending the designer's role of a mere product designer. It offers opportunities to create a new business model in which part of the revenue comes after the first-time purchase. For example, the upgradability approach allows modifications to be made to the garment in the future and this option will be designed into the garment already in the initial stage (e.g. modular design). The garment can be returned to the design studio to be upgraded according to the wearer's changing size, tastes or needs. This could also offer new profitability to a designer or a company.

Some companies already offer repair services and some even take back their old products to use them as patches or as new redesigned products (Niinimäki and Durrani, 2020). This seems to be increasingly common practice in the fashion field and a number of fashion brands now offer repair and even alteration services. Mending no longer has to mean something invisible and modest but can be a creative act or even a statement of sustainability which designers or even consumers can do themselves. In this way, even consumers can take part in the 'design' process by making their own 'mark', or 'design' on garments (Durrani, 2018, 2021; Niinimäki and Durrani, 2020).

In the future a more critical view will be taken of the current way of designing and designers will be able to create opportunities to not only extend the use time of a garment but also make the garment more open and flexible, offering or even encouraging consumers to make alterations according to their own needs. In a way, consumers need to be designed into the product narrative and see their own actions as a sustainable possibility to create meanings in their own material world, extend garment lifetimes and thus slow down consumption.

Designing for recycling; intentional design

In moving towards a circular economy and aiming to close the material loop it is more important than ever to focus on design for recycling strategies. Products need to be designed in a manner that enables them to be dissembled, after which different components can be separated and reused as they are (e.g. zippers, buttons), and then made into new garments or recycled as material. This approach requires a completely new way of designing materials and products that are suitable for disassembly and re-use or suitable for material recycling.

The Design for recycling strategy also includes integrating the after-use phase into the design process. This is a new knowledge area for designers. This approach is also known as *intentional design for recycling* and includes knowledge of reverse logistics, product disassembly, material identification, and material recycling and recovery. In this way, the product lifetime and the end of first-use time is defined already in the product design phase. Material selection and assembly work need to make product recycling possible, and this end-of-life knowledge limits the possibilities in the product design phase (Niinimäki and Karell, 2019). This means that a lot of new information needs to be brought into the design process. The recycling method needs to be decided on already when selecting materials for the product or details in the assembly work.

Mechanical, thermal and chemical recycling methods currently exist in the textile field, some of which are developing rapidly (e.g. scaling new innovations for textile waste chemical recycling). The design for recycling strategy might limit the use of some materials which are harder or impossible to recycle (this will be discussed further in Chapter 9). Furthermore, current material identification technologies are under-developed, which also limits material selection or even fabric structure (this will be discussed further in Chapter 8). 'The idea is to give designers design boundaries for what certain types of recycling system can and cannot tolerate, and through this approach to begin to tailor product design to certain types of recycling processes' (Henninger et al., 2022, p. 94).

The design for recycling strategy needs constant development by using the newest research knowledge in the design process, as many recycling technologies are now progressing quickly. We should understand that the strong push towards textiles circularity will drastically change design processes from this perspective and that this new knowledge will affect design work and design choices, at least in material selection and assembly decisions. On the other hand, this new stage (recycling and material recovery) is an important part of the new product lifetime understanding.

Conclusions

Kretschmer (2014) highlights that design and simultaneously the designer's role has to change as our understanding of sustainability improves. 'The commonly accepted role of design as a cheap resource of ideas, as a profession of beautification, and as a powerful marketing "tool" must therefore be questioned seriously and urgently' (Kretschmer, 2014 p. 183). Designers play a critical role in the aim towards extended product lifetimes as well as in designing and manufacturing textiles and garments that are suitable for recycling. New knowledge and skills are really needed in this area. I end this chapter with Jo Cramer's words (2019, p. 18) on following design guidelines while considering the longevity aspect:

- Design for durability: Reinforce areas of wear, use durable fabrics and trim, robust construction
- Design for adaptability: Allow for changes in size and style
- Design for easy intervention: Use simple production equipment and methods so later changes are easy to perform
- Design for easy replacement of parts: Use a combination of fabrics within garments such that replacement with contrasting fabrics is suitable

- Design opportunities for renewal through re-making
- Design for a range of aptitudes: Consider the skill level of the wearer in making repairs and modifications, offer simple as well as advanced options

These are good principles with which to start. Designers can try their best to produce better quality and offer deeper use satisfaction, but it is ultimately the user who makes the decision to end the lifetime of a product and dispose off the garment. Yet the recycling phase helps extend the lifetime of a material and returns the responsibility to organise reverse logistics and take part on textile recycling to the manufacturers and importers.

We can also consider the lifetime of the garment as something that evolves and consider garments as more open items, which can change in appearance over the years. As McNeill et al. (2020) state 'If consumers are encouraged to see fashion garments as living products, designed to evolve and transform in an extended lifecycle philosophy (Gwilt and Pal, 2017), there is potential to substantially increase the average use of each item.' This approach encourages designers to offer and consumers to adopt more creative ways to use and maintain garments. We can rebuild our relationship with our garments and begin to cherish and love both them and our material world again.

Acknowledgements

This research was supported by the Strategic Research Council at the Research Council of Finland, Grant no 352616 FINIX consortium.

References

Bubonia, J.E. (2014). *Apparel Quality. A Guide to Evaluate Sewn Products.* New York: Fairchild.

Cramer, J. (2019). *Living Wardrobe.* Doctoral dissertation, RMIT, Australia. https:// researchrepository.rmit.edu.au/discovery/delivery/61RMIT_INST:Research Repository/9921863896801341#13248374120001341

Cooper, T., Hill, H., Kinmouth, J. and Townsend, K. (2013). *Design for Longevity: Guidance on increasing the active life of clothing.* WRAP, UK. https://wrap.org. uk/resources/report/design-extending-clothing-life

Cooper, T. and Claxton, S. (2022). Garment failure causes and solutions: Slowing the cycles for circular fashion. *Journal of Cleaner Production,* 351: 131394.

De Klerk, H.M. and Lubbe, S. (2008). Female consumers' evaluation of apparel quality: Exploring the importance of aesthetics. *Journal of Fashion Marketing and Management,* 12(1): 36-50.

Durrani, M. (2018). Designers by any other name: exploring the sociomaterial practices of vernacular garment menders. *In:* Design Research Society International Conference: Catalyst. Vol. 4, pp. 1731-1746. DRS International Conference Series. London: Design Research Society.

Durrani, M. (2019). *Through the Threaded Needle: A Multi-sited Ethnography on the Sociomateriality of Garment Mending Practices.* Doctoral dissertation, Aalto University. https://aaltodoc.aalto.fi/handle/123456789/41489

Durrani, M. (2021). Like stitches to a wound: Fashioning taste in and through garment mending practices. *Journal of Contemporary Ethnography*, 50(6): 775-805.

Durrani, M. and Niinimäki, K. (2023). Color matters: An exploratory study on the role of color on clothing consumption choices. *Clothing Cultures* (forthcoming).

Feather, B.L. (1984). *Men's Wear: Garment Fit, Quality And Care.* Home Economics Guide. University of Missouri-Columbia. https://mospace.umsystem.edu/xmlui/handle/10355/71450

Goldsworthy, K. (2017). The speedcycle: A design-led framework for fast and slow circular fashion lifecycles. *The Design Journal*, 20(sup 1).

Gwilt, A. and Pal, R. (2017). Conditional garment design for longevity. PLATE Conference, 8-10 October, Delft Technical University.

Henninger, C., Niinimäki, K., Jones, C. and Cano, M.B. (2022). *Sustainable Fashion Management.* London, New York: Routledge.

Ikävalko, I. (2022). *The Ill-fitting Clothes: A Study on Size Problems and Their Environmental Impact.* Master Thesis, Aalto University, Finland.

Karell, E. (2014). *Planned Continuity: Design of Sustainable Clothing Service Concept.* Master thesis. Aalto University, Finland. https://aaltodoc.aalto.fi/handle/123456789/13401

Karell, E. and Niinimäki, K. (2020). A mixed-method study of design practices and designers' roles in sustainable-minded clothing companies. *Sustainability*, 12(11): 4680.

Kozlowski, A., Bardecki, M. and Searcy, C. (2012). Environmental impacts in the fashion industry: A life-cycle and stakeholder framework. *Journal of Corporate Citizenship*, 45: 17-36.

Kretschmer, M. (2013). A design perspective on sustainable innovation. *In:* Gassmaan, O. and Schweitzer, F. (Eds.), *Management of the Fuzzy Front End of Innovation*. pp. 179-191. New York: Springer.

Laitala, K.M., Boks, C. and Klepp, I.G. (2015). Making clothing last: A design approach for reducing the environmental impacts. *International Journal of Design*, 9(2): 93-107.

Lilley, D., Bridgens, B., Davies, A. and Holstov, A. (2019). Ageing (dis)gracefully: Enabling designers to understand material change. *Journal of Cleaner Production*, 220: 417-430.

McNeill, L.S., , Hamlin, R., McQueen, R.H., Degenstein, L., Wakes, S., Garrett, T.C. and Dunn, L. (2020). Waste not want not: Behavioural intentions toward garment life extension practices, the role of damage, brand and cost on textile disposal. *Journal of Cleaner Production*, 260: 121026.

Morseletto, P. (2020). Targets for a circular economy. *Resources, Conservation and Recycling*, 153: 104553.

Niinimäki, K. (2014). Green aesthetics in clothing: Normative beauty in commodities. *Artifact*, 3(3): 3.1-3.13.

Niinimäki, K. (2015). Consumer behavior in the fashion field. *In:* Muthu, S.S. (Ed.), *Handbook of Sustainable Apparel Production*. pp. 271-287. Taylor and Francis: CRC Press.

Niinimäki, K. (2018). Knowing better, but behaving emotionally: Strong emotional undertones in fast fashion consumption. *In:* Becker-Leifhold, C. and Heuer, M. (Eds.), *Eco Friendly and Fair: Fast Fashion and Consumer Behavior*. pp. 49-57. London, UK: Routledge, Taylor and Francis.

Niinimäki, K. and Koskinen, I. (2011). I love this dress, it makes me feel beautiful: Emotional knowledge in sustainable design. *Design Journal*, 14(2): 165-186.

Niinimäki, K. and Armstrong, C. (2013). From pleasure in use to preservation of meaningful memories: A closer look at the sustainability of clothing via longevity and attachment. *International Journal of Fashion Design, Technology and Education*, 6(3): 190-199.

Niinimäki, K. and Karell, E. (2019). Closing the loop: Intentional fashion design defined by recycling technologies. *In:* Vignali, G., Reid, L.F., Ryding, D. and Henninger, C. (Eds.), *Technology-driven Sustainability: Innovation in the Fashion Supply-chain*. pp. 7-27. Cham, Switzerland: Palgrave Macmillan.

Niinimäki, K. and Durrani, M. (2020). Repairing fashion cultures: From disposable to repairable. *In:* McNeill, L. (Ed.), *Transitioning to Responsible Consumption and Production*. pp. 154-168. Basel Switzerland: MDPI Publication, https://www.mdpi.com/books/pdfview/edition/1246

Verbič, T. (2019). *Re-designing Classic Wardrobe Items*. Master thesis, Aalto University, Finland. https://aaltodoc.aalto.fi/handle/123456789/37555

Fashion Designers Leading the Way to Extend Textile Lifetimes in Public-private Enterprises

Ulla Raebuild* and Vibeke Riisberg
Design School Kolding, Denmark
e-mail: *ur@dskd.dk

Introduction

How might fashion and textile designers enable change towards green transition in society? As studies have shown, change at product level alone will not bring about the necessary results needed to help prevent climate change (see e.g., Global Fashion Agenda, 2017; Niinimäki et al., 2020). Therefore, the design community must seek out other avenues for the systemic changes that are needed. In this chapter we investigate the notion of *design leadership* as one such avenue. Design leadership is understood as taking on new roles to identify the root causes of unsustainability and taking actions to change the situation from a systemic and holistic world view (Burns et al. 2015; Rissanen, 2018; Benmira & Agboola, 2021). This is of interest because it might be a way for designers to close the gap between general existing understandings of sustainability in the fashion industry: management focus on lowering costs and greenhouse gas emissions and designer focus on building new values and cultures (Thomas, 2020). Together, they can be seen to represent economic, environmental, social, and cultural aspects of sustainability, also known as the four pillars of sustainability (www.fashionseeds.org). However, designers can find it difficult to embed cultural and social aspects of sustainability into their practice in the conventional fashion industry. We therefore agree with Fletcher & Grose that a key question of "how to relate to sustainability in design practice involves uncovering potential in contexts beyond the norm", as alternative sectors in the economy "provide more opportunities for designers to apply their professional skills to the public and ecological good" (Fletcher & Grose, 2012, p. 155).

In the chapter, we will examine two examples of what can happen when fashion designers leave behind the context of conventional fashion and venture into new parts of society across the public-private sectors. We aim to show and understand, how these new arenas provide spaces for them to practice and build new designer roles and how those engagements might foster leadership within other systems apart from the global fashion industry? Lastly, we aim to shed light on how fashion designers can lead the way to actual change – in this case for extended textile lifetimes. The field of fashion and textiles are inextricably linked and in the following we take that as a prerequisite.

As a background, we first address proposals for new designer roles in fashion and textiles. We then introduce the key understanding of leadership that we will apply in the text. Next, we briefly touch upon circular systems of relevance to the chapter. This is followed by a deep dive into two practice examples of designer engagement in Danish public-private enterprises with a focus on extending textile lifetimes in different settings. We use these to elucidate, exemplify and discuss the ways in which fashion design leadership might be identified and seen to occur, including perspectives on further sustaining fashion design leadership for extended textile lifetimes in industry and education.

Key concepts: New designer roles and leadership for systemic change

New roles

The discussions of design practice that currently takes place as a reaction to unsustainable systems leads to the question of designer roles. If designers engage in not only design of materials and products, but also design of new services and systems for circularity and product lifetime extension, then what might be the new roles they need to take on in industry and society at large, to enable the development of those new services and systems? In their visionary book *Fashion Design & Sustainability: Design for Change* (2012), Fletcher and Grose proposed four new ways of operating: (1) *Designer as communicator-educator* is to "take abstract information, which is often ineffective at prompting action, and make it real and appropriate, to trigger new behaviour" (p. 158). (2) *Designer as facilitator* is to be "comfortable with the unknown, synthesizing complex information, working across disciplinary boundaries and making intuitive leaps in thinking (Fletcher & Grose, 2012, p. 162). (3) *Designer as entrepreneur* is to "ask how new businesses will be built and how they will differ from what has gone before; what new roles design will play in them; and what sort of aesthetic will emerge when the products and services of the fashion sector are built on a fundamentally different set of values" (ibid., 2012 p. 174), and (4) *Designer as activist* is "combining the contribution of both public and private sectors, in actively pursuing design opportunities that span civil society as well as government and the commercial institutions of the market" (ibid. p. 169).

These proposals for new roles in fashion design has been part of Design School Kolding (DSKD) students' mandatory reading since it was published and still holds relevance now ten years after. In fact, we have seen in our work as design educators at DSKD that there is a strong and growing urge in design students to take on these new roles and combine them with conventional ones (Leerberg et al., 2010; Ræbild and Riisberg, 2021). Moreover, it has been suggested, that new roles for the fashion designer provides an opportunity "to rethink the term leadership unburdened by ideas of extant hierarchies, particularly in connection with management" (Rissanen, 2018, p. 3). Thus, in the following, we investigate possible new roles in organizations and systems.

Leadership

Adopting new roles, we will argue, also implies for the designer to take on leadership in various ways depending on the context, and one's place in the organisation or company. In the following, we seek to position fashion design leadership for sustainability in relation to contemporary theories which we later apply in the discussion of our examples illustrating new roles taken on by two fashion designers.

A recent literature review on the evolution of leadership theories by Benmira and Agboola (2021) concludes that there is no universal definition of leadership and that the theories constantly develop and change according to our complex world. Their study investigates the notion of leadership in general and for our purpose the theories, especially after 2000 are relevant. Among these are *inclusive leadership* which they describe as "a person-to-person approach based on a dynamic process between leaders and followers" (Benmira & Agboola, 2021, p. 4); here the main point is to empower followers to become leaders. Another theory concerns *complexity leadership*, which is developed to deal with our modern globalized world. This approach "takes a whole-system view, considering contextual interactions that occur across an entire social system" (ibid.); however, this study does not mention sustainability. In contrast, theoretical foundations in leadership for sustainability, has been described by Burns et al. (2015). They observe that *Leadership for sustainability* is a relatively new idea which radically expands the earlier notions of leadership. Leaning on Ferdig (2007), they find that this new understanding "includes an enlarged base of everyday leaders in all walks of life who take up power and engage in actions with others to make a sustainable difference in organizations and communities" (Burns et al., 2015, p. 89). Furthermore, the authors underline that *Leadership for sustainability* is not something held by an individual but "demonstrated by people and teams throughout an organization in responsive processes" (ibid., p. 91, citing Kouzes & Posner, 2012). In this sense, leadership becomes an inclusive, collaborative, and reflective practice based on sustainability values rooted in "a living processes paradigm that require dynamic interactions among stakeholders" (ibid.).

From a design practice perspective West (2017) suggests rethinking our future through design leadership and Rissanen (2018) discuss fashion designer's new roles in relation to sustainability, citizenship, and new understandings of leadership. As he points out, this is not to be mixed up with the notion of management, which deals with tasks such as staffing, budgeting, and planning. Leadership, on the other hand is about vision, behaviour and bringing useful change, and therefore not limited to a particular part of an organizational hierarchy (Kotter, 2013; Rissanen, 2018). Inspired by Ruppert-Stroescu & Hawley (2014), Rissanen introduces the two notions, *Leadership Creativity* and *Adaptive Creativity*. The first can overrule current archetypes and shift the fashion sector in new directions, while the second can only contribute to the existing paradigm. In conclusion, Rissanen states that bold solutions to complex systemic problems of unsustainability calls for Leadership Creativity (Rissanen, 2018).

Systems and sustainability

To bring the idea of fashion design leadership into the context of recycling and lifetime management in the textile and fashion sector, it is important to bear in mind the strategies and systems that designers and companies may operate within. For this purpose, we lean on *The Four Models of DCE*, Design in a Circular Economy (RSA, 2016, in Niinimäki, 2018) containing four levels of systemic design-based loops: (1) Design for product longevity, (2) Design for service, (3) Design for re-use in manufacture, and (4) Design for material recovery. Each loop involves different actors: Users, brands, factories, resources and recycling facilities. Design for material recovery (the 4th level) is the most challenging, however, intensive research is carried out to find ways of recycling textile waste into new fibers (ibid, p. 18). To bring about substantial change in terms of resource recovery, these new technologies must be implemented in large scale commercial systems. *The Four Models of DCE* is one of many useful tools which can assist companies to plan new strategies and uncover the systems needed to develop a circular economy (CE). One prominent initiative to promote CE is Ellen McArthur Foundation (www.ellenmacarthurfoundation.org). This NGO is committed to build and share knowledge to redesign the future of fashion systems, by providing open-source tools. However, few fashion companies (if any) have yet been able to implement full CE, mainly due to the complexity of global supply chains, lack of systems to collect and process textile waste. Furthermore, new business models based on e.g., product service systems (PSS) are necessary to control material flows and close the loops. Here, PSS represents an alternative to conventional business models because it is a combination of tangible products and intangible services designed to 'fulfill the customer's needs rather than selling a product'. Furthermore, Tukker and Tischner (2006) consider Product Service Systems to be one of the most promising solutions to the current problem of resource drain and over-consumption (Petersen & Riisberg, 2017, p. 5).

Outside the traditional fashion sector, we find pioneers of PSS such as Patagonia and Vigga (Niinimäki, 2018; Petersen & Riisberg, 2017), and within the Danish public healthcare sector, uniforms and other textile products are handled within such systems (Petersen & Riisberg, 2016). The examples below also provide system models created by two designers (Figure 1), the context and flows they operate in. The models illustrate how fashion designers can approach various system levels and lead the way to extend textiles lifetimes towards circular economy.

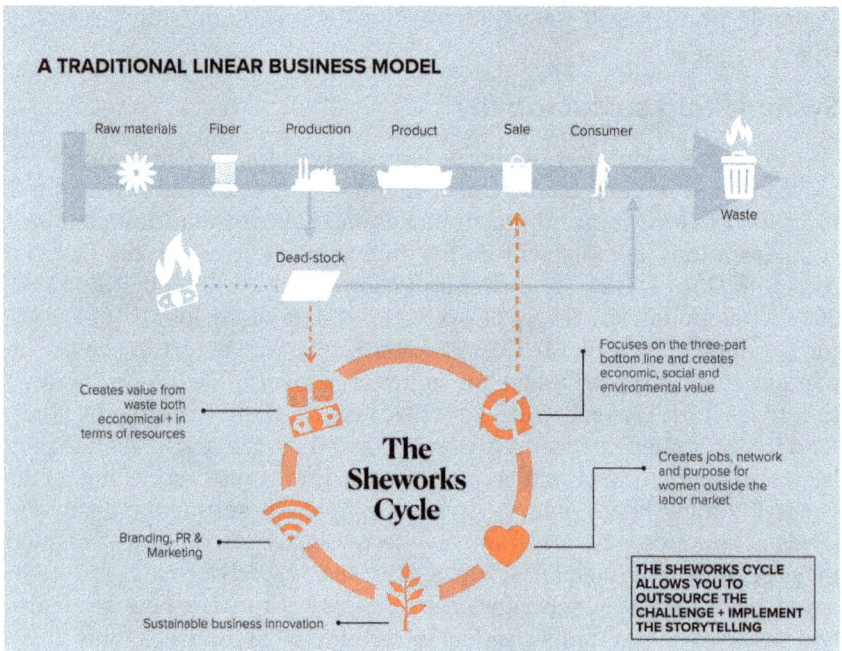

Figure 1. Sheworks systems cycle and business model (Solveig Søndergaard, 2022)

Two designer-enterprise examples

This section describes two designer-enterprise examples, Sheworks and midtVask, through which we examine the new roles in fashion design leadership. The examples share some commonalities: The designers have both graduated from Design School Kolding, however, with some years apart and with different background. The companies they work at are both, to a different extent, public-private enterprises with social inclusion as a key element in their business model and employment strategy. Both companies aim to prolong textile lifetimes through design strategies. In addition, neither of the companies are conventional fashion brands. In that respect both designers draw on interdisciplinary and design for sustainability competencies obtained during various courses at DSKD in combination with traditional fashion and textile skills. The examples differ in that the first, Sheworks, is a designer-initiated start-up, whereas the second, midtVask, was an established company before the designer came on board. This company operates as a product service system, leasing textile products to the hospitals.

To describe the examples and understand the ways in which the designers work in the specific company contexts, we conducted semi-structured interviews (Kvale & Brinkmann, 2009) with the designers and utilised the companies' websites. For background, we further leaned on authors' former site visits, a recent radio podcast with Solveig Søndergaard (Tanggaard, 2022) and Sarafina Taudals MA thesis work (Andersen, 2021). As we have been teaching and supervising both designers during their studies at DSKD, we acknowledge the risk of bias. On the other hand, it is a unique position to be able to follow design students into their professional careers.

Designer Solveig Søndergaard and Sheworks Atelier

Solveig Søndergaard is CEO and head of creative direction at Sheworks Atelier. Søndergaard calls herself a social impacter for environment and society and aims to tackle resource waste in the textile industry in combination with integration and job creation for meaningful lives (see Figure 1) On their website, Sheworks is described as "a Circular Design Studio which utilizes surplus and waste materials from the textile industry in Denmark creating beautiful circular products made by hand" making "circular novelties" (www.sheworks.dk). The studio is situated in Kolding, Denmark and currently employs eight women from Syria, Iran, Iraq, Eritrea, Albania, Czech Republic, Somalia, Germany, and Denmark. About a year ago a fellow designer from DSKD joined and now co-leads design and production with Søndergard.

Søndergaard graduated from DSKD in 2017 with a mixed MA degree in fashion and communication design. Her work during studies gravitated towards communities, craftivism, upcycling, and participatory design, often with public institutions and organisations as key partners. At the tail end of her education, she was employed as research assistant in a bigger research project called *THREAD: Textile Hub for Refugees Empowerment, Employment and Entrepreneurship Advancement in Denmark – A new model*, led by University of Copenhagen with DSKD as partner (2017-2019). In the autumn 2018, Søndergaard began to host weekly craft get-togethers at DSKD for refugee and immigrant women in the area but also open for students and staff at DSKD. The craft products made by the women were sold locally and proved to be of commercial value. From this initiative and after the many dialogues with the local municipality during the research project, grew the idea for Sheworks Atelier.

Søndergaard explains that she was looking for a way to translate theory into practice and therefore started the workshops. Here she grew a network of women involved in the project. She saw how the women held competencies but were overlooked in the Danish systems. For example, the women were skilled in making embroidered, knitted, and crocheted products but the money they earned by selling these locally, was deducted from their allowance. Thus, there was no incentive to support themselves in this way. Søndergaard felt this was unfair to the women and realized that the only way forward was to make a company, where the women could get proper jobs. After some months of negotiations with Kolding municipality, she managed to convince them to collaborate in a pilot project, where they together tested if the women where interested in joining a startup company. Clearly, the Municipality should have a strong interest in such an endeavor, as the support of a single parent refugee is around 1.7 million Danish Kroner pr. year, so there was something to gain both in human and economic terms. However, the inertia of the public system is not geared to react quickly due to rules and regulations. Luckily, they were open to the idea and Søndergaard's convincing arguments were supported by the results of the Thread research project. The pilot turned out successful and when it ended, the municipality withdrew as an economic partner, and the company became self-financed. However, they learned that this way of engaging with a designer created a productive synergy for all stakeholders and just decided to begin a new collaboration for people of Danish ethnic origin outside the labour market. Søndergaard points out that the employed women use each other as 'support system' and explains that the feeling of being good at something 'lights a fire within them'. She experiences the craft processes as helpful for the employees, in that the creative engagement provides a space away from what they are dealing with privately.

In terms of designing products from industry textile waste, Søndergaard explains that the process is about merging diverse cultural craft techniques with a Danish Nordic aesthetic for the markets they operate in. The women make prototype proposals which are then off set for a dialogue on commercial potentials and adjustments in design, which generate a lot of interesting conversations and mutual learning on taste. Søndergaard recalls, for example, how one of the ladies' described the Danish taste in colour as 'the colour of dead animals'. She tells how they have made a 'competence library' which they use to atune their product designs to different markets aesthetic preferenes, e.g. Scandinavia, Holland, and Japan.

The product range includes decorative pillows, bags and wall decorations which is sold online to private customers and to companies (B2B). Their design approach is material driven and zero-waste. New products are developed continuously depending on the types of pre-consumer textile waste they receive from collaborators in the Danish fashion and textile industry (examples are Ganni, Georg Jensen and YKK). They re-use approximately 708 kg of textile waste per year and welcome both bigger and smaller orders, as stock also varies. Up to now they have experienced that their costumers welcome the inherent variation of the designs, as it provides a bespoke quality where very few products are completely alike. So far, they haven't been short of supply, but they have high standards in terms of quality of textiles and thus seek high-end collaborators. They do not accept low-quality fashion textiles nor post-consumer textiles. They also reuse high quality leather.

Søndergaard reflects on the importance of meaning in her work, and says, that for her, as a designer, making beautiful things is no longer enough; it feels hollow. She believes her generation seeks impact enterprises where the product is the means to the goal. It is not about the decorative cushion, but about the change it can foster personally and socioeconomically. She states that her work needs to have a systemic effect and point out that tools from fashion and design are useful. Søndergaard acknowledges, that to be able to make a difference may sound clichéd, but the aim is also to set a new societal agenda.

Søndergaard says, that she knew it would be difficult, anxiety provoking and would demand a lot of courage and power to become an entrepreneur. But as a person, she doesn't mind taking risks and likes to solve complex problems. Her motto is 'stay stupid', which to her means keep being curious, willing to learn and reach out to others. She finds that designers are good at analyzing and recognizing models and systems, decoding patterns and trained to be in uncertain processes. To illustrate this, she uses the analogy: to walk around the mountain and look down at where you were before and then copy what you see into new solutions.

Designer Sarafina Taudal at midtVask

Sarafina Liv Taudal is Sustainability Consultant and Designer at the public-private hospital laundry service midtVask. Taudal graduated from the Design for Planet MA in design and sustainability at DSKD in 2021. She has a BA in fashion design from Via University College – an industry focused profession education. After her BA, she worked a couple of years for a large fast fashion brand in Denmark.

midtVask is a company providing a Product Service System based on a laundry and repair service for hospitals. They own all the textile products (bedlinen, towels, uniforms etc.) and therefore able to control the life cycle including development of products for a circular system (see Figure 2). midtVask services approximately 10 hospitals in the middle Region of Jutland, Denmark, where a staff of 160 people maintain 28 tons of textiles per day – uniforms, bedlinen etc. The company is a socio-economic enterprise that educates and employs refugees as well as Danish citizens with social challenges. The staff currently represent 36 different nationalities, that rotate between different workstations and functions during the day, including manual work for the sewing workshop where recycling and repair has increased the workload.

Taudal worked at midtVask during her MA studies. The initial task was to address upcycling of textile waste, but early on she initiated a broader focus from only re-using textile waste, to also out phasing waste. After graduation, midtVask employed her for seven months in a new project position, supported by the regional Centre for Sustainable Hospitals, which then lead to the permanent full-time position she currently holds.

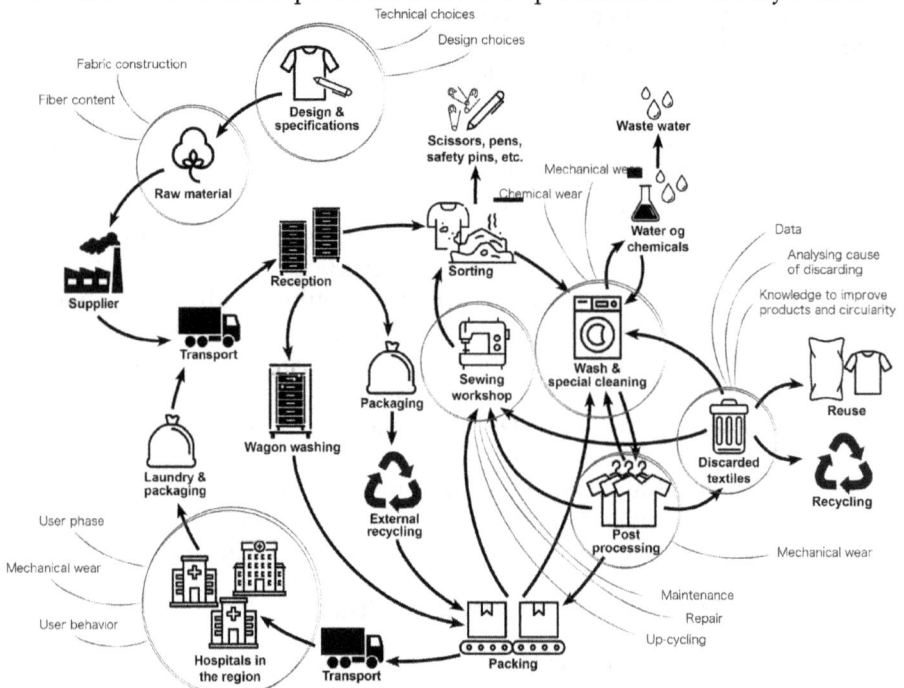

Figure 2. midtVask system and points of design intervention model
(Sarafina Taudal, 2021)

Taudal started out doing work on product demand specifications and technical packages for production, but soon she also got engaged in various innovation projects. All along, she has worked closely with a colleague who knows the organization well, to apply her knowledge about design, sustainability, and innovation. Taudal first mapped out all the processes that the textiles go through in terms of production, use and afterlife to understand where the company should intervene and place their efforts. She identified the following points for intervention: the use phase at the hospitals, the laundry machinery and processes at midtVask and issues to do with the original design. For instance: Can the item be repaired? Is it technically strong enough for the laundry process? Furthermore, she found buyers for the discarded textiles leading to reduction of textiles' waste at midtVask.

A lot is happening right now, where she is systematizing everything to train someone else to take over some of the tasks. To change the current system, they need to have valid data on where the waste originates from. Taudal also prepares upcoming procurements and ad hoc tasks such as a new birth pillow commissioned by a birth ward, including the design, technical development, and price. She has also initiated a new practice where she oversees the contact with suppliers and describes the product demand specifications, because she knows about textile qualities and reinforcements that are important technical requirements for cyclability.

Taudal explains that she decides together with nearest leader and head of company, which projects, and initiatives they enter whether for internal or external application. Internally they have begun to repair and re-manufacture for their own extensive stock. For example, they remake sheets into 'multi-sheets' (a smaller kind of sheet), so they won't have to be produced from new material. In similar ways, remaking adult duvets into junior duvets for the Aarhus Municipality. They also recirculate out-phased products. Taudal gives the example of panties used in the Emergency department, where thousands of panties disappear every year because clients take them home. Now they use out phased panties to hand out, instead of new ones.

Because they are a non-profit enterprise, they are not allowed to donate any profits to charity. But they do donate hospital work wear to Ukraine, while making sure that the material is fully functioning and not a way of getting rid of waste. midtVask is also involved in several projects on fibre-to-fibre re-use, but Taudal estimates that the technology is still some time away, in terms of implementation.

In the nearly four years that Taudal has been at the company, she sees some clear results. Although they need better data collection, the estimated waste prior to the new circular initiatives was 48 tons pr year, which is now estimated to be only 18 tons pr year. She acknowledges that compared to the 28 tons laundered every day, it might not seem as much, but to her, it is resources saved. However, there is still a huge unaccounted wastage during use.

Taudal says that her motivation is "to contribute towards making and doing things better". There hasn't been a lot of focus on textiles in the public sector and resources can be used better. She sees so many 'low hanging fruits. To her, sustainability work is not a particular branch of an organization, but part of the core that must be incorporated into company operations. She sees that there is still some way to go, but that there is a significant shift in mindset of the people she works closely with. Others still talk about "this sustainability thingy". She explains that there are a lot of engineers in the management team as well as production managers who understand sustainability differently. "Engineers talk about data, production leaders about practice, what it means for their daily practices on the floor." She has had to learn to communicate to a motley crew, and say yes, it might be slightly more bothersome, but what do we gain?

Taudal emphasizes that she welcomes national and EU legislation. To her, it provides opportunities more than limitations. There are so many new actors that wants to join in now, compared to two years ago, where no one was interested. She sees that textile waste is now seen as a resource and they can begin to build revenue from their own textile waste. It starts to pay off to sort their waste. However, she is adamant that the stories the company communicates externally need to be valid. Stories are important but they must steer clear of green washing.

She explains, that at times it is a barrier to be a public enterprise, as they have a particular economic framework in terms of procurement. Some products can't be altered for the better before a new round of procurement. They must wait. Also, there are many requirements when you work in the hospital sector. Taudal says, that she addresses these constraints as she would in any design process, which is fruitful for her creativity. She never liked starting from a blank piece of paper. She likes working around functionalities, but "it is the kind of work that requires patience, as the effects are measured in the long run".

Next up, Taudal says she would like to be able to develop a proper strategy, so they can work less on assumptions of what they want and do, and more towards goals and how to reach them. To her, it would be beneficial and motivating to then achieve these goals and set new ones. "To feel the victories, even when they are small."

The shift from private fashion company to public textiles enterprise has been huge. But Taudal says, that it makes much more sense contributing to a better world "to put it on the grand scale". In her former work, she was often conflicted within herself. It was just another dress and all about making money. She elaborates: "Where I work now, it is also about the bottom line, but other parameters are also important. I can draw on all my competencies from fashion when I for example have to design a new polo from circular principles, but I do not miss working with e.g., mood boards. I use my design skills in a different way, to bridge the theoretical with the hands on. Making all kinds of things for the company from signs and presentations to ways of engaging colleagues in saving textiles. I sometimes say it is an unsexy job, but the meaningful aspects are most important."

Discussion and further perspectives

We began the chapter by asking: what happens when fashion designers leave behind the context of conventional fashion and venture into new parts of society a cross the public-private sectors?

To answer our question, we have analysed interview data in relation to Fletcher and Grose's (2012) description of new designer roles and in the following we use the outcome to discuss possibilities concerning leadership and systems change for extended textiles lifetimes.

The two examples demonstrate, that both designers have developed systems from waste thus extending textile lifetimes for the benefit of the environment. Furthermore, they are both engaged in creating cultural, social, and economic value through different business opportunities based on these extensions. Both take part in designing products that add loops to existing systems, teach staff to operate and co-develop for and within these. They also both work with unknown and complex factors such as material flows and legislation across public private sectors.

However, there are also significant differences in the practices when it comes to scale and type of the system, size of textile waste volumes, product ranges and types of engagements internally and externally. As mentioned before, midtVask operates as a product service system and therefore they can reuse part of the textiles for new products and purposes and extend the lifetime into new cycles within their own organisation. Sheworks, on the other hand is a small upcycling enterprise, unable to control products post-purchase, meaning that they can only design for one loop to extend lifetime. When it comes to new designer roles, there is also a difference between the two examples.

Looking at Søndergaard, we see a clear profile for *designer as entrepreneur* in the way she builds a new socio-economic business around design, waste from the textile industry and employing the women's craft skills to create social and economic value. She also displays *designer as activist* traits in her fundamental motivation to address the inertia of the municipality system, create alternative jobs, inclusion, and social change through her role as 'social impactor' (in her own words). However, we also identify strong elements of *designer as educator/communicator* and *designer as facilitator* in her engagement with teaching employees, facilitating workshops, and communicating Sheworks values and mission at all levels, in different media and across all stakeholders.

Looking at Taudal, we see a clear profile for *designer as facilitator* in the way she identifies points of intervention in the current system to facilitate extended textiles lifetime to a new level in the organization. As a facilitator she uses her design competencies to bridge disciplinary boundaries and meet the many requirements in the hospital sector. This role is closely connected to *designer as educator/communicator* in the way she works with the staff, management, and external stakeholders. As Søndergaard, Taudal also engages in dissemination at e.g. conferences, seminars and within design education. She also carries with her an awareness of new business opportunities that arise from within the altered system and seeks to put them in motion through her design competencies. Thus, a strand of *designer as entrepreneur* is at play.

The examples thus demonstrate how Fletcher and Grose's notions of new roles are present and visibly intertwined in the designer's practices. We will now address how leadership might be seen to consequently emerge. For this purpose, we suggest seeing the notions of leadership introduced in the chapter as representing a general-, a sustainability- and a fashion and textile design perspective on leadership.

From a general leadership perspective both designers in various ways practice a person-to-person approach, also known as *Inclusive leadership,* that empower followers to become leaders. This is apparent in the way they train and guide people in the organization across hierarchies and can adjust their engagements according to audience/person. Furthermore, both display *Complexity leadership* in the way they take a whole-system view (Benmira & Agboola, 2021) in their practice exemplified in their systems models (Figure 1). They also display thorough understanding of the importance to assess effects within the complete system they work within, as well as the bigger systems their organization is part of.

From a sustainability perspective, we recall that *Leadership for sustainability* is characterized as an inclusive, collaborative, and reflective practice based on personal values and beliefs (Burns et al., 2015). We see that both designers describe strong motivations for making a difference for people and planet. Moreover, *Leadership for sustainability* requires dynamic interactions among stakeholders, this is manifested in the designers' actions across the public private sectors despite many challenges. It is also visible in e.g., the collaborative design practice at Sheworks that Søndergaard has developed and that has led to a unique selling point by merging various cultural aesthetic elements. Both identify problems in systems, and act upon them based on a reflective design practice that builds on data, scientific and hands on knowledge.

Finally, from the perspective of fashion and textile design we find that Rissanen's point of *Leadership Creativity* (Rissanen, 2018) as a way to overrule current archetypes in the industry and shift the fashion (here textile/garment) sector in new directions can be applied to the examples. Both designers use their creativity and core designer skills in full but seek to apply these in ways that can redirect design practice for new purposes, here for the purpose of extending textile lifetimes (as opposed to the product obsolescence in the conventional linear fashion system). They thus showcase how fashion designers can build careers in sectors other than the conventional fashion industry, and by doing so, can begin to slowly make change happen and possibly influence the industry they have abandoned. However, to some extend Taudal similarly performs *Adaptive Creativity* as her work needs to contribute to an already existing organisation. Since Søndergaard and Taudal operate outside the conventional fashion industry, they do not affect that system directly, but by setting examples for other designers and students they lead the way into new roles suggesting alternative routes that can be applied also to fashion production. In fact, they are part of a global movement where designers find alternative ways of practicing fashion design skills in so many new ways witnessed in the huge number of publications on fashion design for sustainability and NGOs like Fashion Revolution (www.fashionrevolution.org). Thus, overall, we see in the examples the designers practice a more or less articulated form of leadership *as vision, behaviour and bringing useful change* (Rissanen, 2018) not limited to a particular part of an organizational hierarchy.

In terms of systemic impact and actual change for extended textile lifetime, the two examples display various steps on the way towards resource recovery and circularity. As shown in the Sheworks model (Figure 1), only one loop is added which extends textile lifetime, this relates to level 1: Design for product longevity, in *The Four Models of DCE*. Thus, the impact of the enterprise can be seen as primarily raising awareness of deadstock textiles as a valuable resource that should not be wasted. However, Sheworks also has an impact and creates change as a 'role model' by implementing social, economic, and environmental sustainability with value for the employees and local society.

At midtVask, recovery of textiles for extended lifetime is at a different scale and takes place in several loops. Thus, creating systemic impact at the three first design-based loops of *The Four Models of DCE*, (1) Design for product longevity, (2) Design for service, (3) Design for re-use in manufacture. As stated by Taudal in the interview (4) Design for material recovery, is not yet in place, but the company participates in a project concerning large scale fiber recycling as they aim to include this cycle in their strategy going forward.

This study has uncovered ways in which the exemplar designers incorporate notions of leadership in their design practice towards extended textile lifetimes, how design thinking, practice and leadership for sustainability are utterly entangled, and furthermore how new roles grow leadership. As a further perspective we propose and advocate the potential to bring out leadership as a notion in education in alignment with design thinking and practice for sustainability.

References

Andersen, S.L.T. (2021). *A Design Tool towards a more Sustainable Fashion Practice.* Design for Planet MA Thesis, internal document. Design School Kolding: Kolding, Denmark.

Benmira, S. and Agboola, M. (2021). Evolution of leadership theory. *BMJ Leader*, 5: 3-5.

Burns, H., Diamond-Vaught, H. and Bauman, C. (2015). Leadership for sustainability: Theoretical foundations and pedagogical practices that foster change. *International Journal of Leadership Studies*, 9(1): 88-100.

Ellen MacArthur Foundation (online): www.ellenmacarthurfoundation.org, 04/08/2022

FashionSEEDS (online): https://fashionseeds.org, 30/05/2022

Ferdig, M.A. (2007). Sustainability leadership: Co-creating a sustainable future. *Journal of Change Management*, 7(1): 25-35.

Fletcher, K. and Grose, L. (2012). *Fashion Design & Sustainability: Design for Change.* London, UK: Laurence King Publishing.

Global Fashion Agenda & Boston Consulting Group (2017). *Pulse of the Fashion Industry.* Copenhagen, Denmark: GFA.

Kotter, J.P. (2013). Management is (still) not leadership. *Harvard Business Review* (online): https://hbr.org/2013/01/management-is-still-not-leadership, 04/08/2022

Kouzes, J. and Posner, B. (2012). *The Leadership Challenge* (5th edition). San Francisco: Jossey-Bass.

Kvale, S. and Brinkmann, S. (2009). *InterViews: Learning the Craft of Qualitative Research Interviewing.* California, US: Sage Publications.

Leerberg, M., Riisberg, V. and Boutrup, J. (2010). Design responsibility and sustainable design as reflective practice: An educational challenge. *Sustainable Development*, 18(5): 306-317.

midtVask (online): https://midtvask.dk 16/03/2022

Niinimäki, K. (2018). Sustainable fashion in a circular economy. *In:* Niinimäki, K. (Ed.), *Sustainable Fashion in a Circular Economy.* Helsinki, Finland: Aalto University School of Arts, Design and Architecture.

Niinimäki, K., Dahlbo, H., Peters, G. and Perry, P. (2020). The environmental price of fast fashion. *Nature Reviews, Earth & Environment*, 1: 189-200.

Petersen, T. and Riisberg, V. (2016). Pockets, buttons and hangers: Designing a new uniform for health care professionals. *Design Issues*, 32(1): 60-71.

Petersen, T. and Riisberg, V. (2017). Cultivating user-ship? Developing a circular system for the acquisition and use of baby clothing, *Fashion Practice*, 9(2): 234-254.

Riisberg, V. (2019). A spotlight on: VIGGA.US – Sharing baby clothes in a sustainable product service system. *In:* Gwilt, A., Payne, A. and Ruthschilling, E.A. (Eds), *Global Perspectives on Sustainable Fashion.* pp. 110-112. UK, London: Bloomsberg.

Rissanen, T. (2018). Fashion Design Education as Leadership Development. Proceedings of Global Fashion Conference, London College of Fashion, October 31st-November 1st, 2018: London, UK.

Rissanen, T., Grose, L. and Riisberg, V. (2018). Designing Garments with Evolving Aesthetics in Emergent Systems. Proceedings of Global Fashion Conference, London College of Fashion, October 31st-November 1st, 2018: London, UK.

RSA Action and Research Centre (2016). *Designing for a Circular Economy: Lessons from the Great Recovery 2012-2016.* London, UK.

Ruppert-Stroescu, M. and Hawley, J.M. (2014). A typology of creativity in fashion design and development. *Fashion Practice*, 6(1): 9-35.

Ræbild, U. and Riisberg, V. (2021). How to design out obsolescence in fashion? Exploring wardrobe methods as strategy in design education. Proceedings of Product Lifetime and the Environment 2021, 26-28 May 2021: Limerick, Ireland.

Ræbild, U. and Riisberg, V. (2021). In the realm of the social: New approaches to fashion and textile design, *In:* Bertola, P. (Ed.), *Fashioning Social Innovation: Design Empowering Communities to Foster Sustainability in Culture Intensive Industries,* Milano, Italy: Mandragora.

Sheworks (online): https://sheworks.dk : 16/03/2022

Tanggaard, L. (2022) (online): Danish National Radio, https://www.dr.dk/lyd/p2/pa-vaerkstedet/pa-vaerkstedet 24?fbclid=IwAR21kn8nFgtQ8bwBafZb19KTDlOm59Iz2gqAEUz3vMsZRA4-9YudnEcYs_o : 04/08/2022

Thomas, K. (2019). Cultures of sustainability in the fashion industry. *Fashion Theory*, 24(5): 1-28.

Tukker, A. and Tischner, U. (2006). Product-services as a research field: past, present and future: Reflections from a decade of research. *Journal of Cleaner Production*, 14(17): 1552-1556.

West, L. (2017). Design Leadership: Rethinking Our Futures (Part 1), https://lucy-west.medium.com/design-leadership-rethinking-our-futures-162520843c46, accessed 22.6.23

I Like It because It's been Worn Before: The Sensory Longevity of Worn Clothing

Mila Burcikova

University of the Arts London, UK
e-mail: m.burcikova@arts.ac.uk

Introduction

This chapter highlights that designing for extended clothing lifetimes should include a focused creative reflection on user's multi-sensory perceptions of clothing in long-term use – introduced here as sensory longevity of clothing. As Bridgens and Lilley (2017) note, while designers are producing objects that are to be used in the future, they rarely look at that future. However, if clothes are to be worn over extended periods of time, learning to project long-term use must be at the core of the design process. The research presented here highlights that while the current fashion system is geared towards an obsession with newness and constant change, women's relationships with clothes in their wardrobes can unfold slowly, through multi-faceted considerations and sensory perceptions that often favour familiarity acquired through long-term use over the push to look for new and supposedly better, choices. In short, as Skjold (2016) too points out, the logic of current fashion business does not seem to reflect the "logics and practices of the majority of people when they go about getting dressed every day" (p. 136), with devastating impacts on both the planet and people.

The argument presented here builds on my doctoral research that investigated emotional durability of clothing through the lens of my designer-maker practice. The approach I chose considered especially how women's sensory engagement with their clothes figures in their relationships with their wardrobes and how these relationships could inform the creative practice of designers who wish to design for continuity

(Skjold, 2014) and increased user satisfaction (Niinimäki, 2014). The research journey I chose was shaped by Sarah Pink's work on sensory ethnography (2015 [2009]), which led me to focus especially on women's hands and the ways in which clothes were handled. My research approach also reflected Norman's (2004) argument that look and feel in perception of design (*visceral level*) often precede considerations of how things work (*behavioural level*) and how they link to our self-image and memories (*reflective level*).

This chapter builds on the relationship of sensory comfort and construction details to clothing longevity, as these have received little consideration in research to date. For an extended discussion of sensory experiences, and other contextual aspects of emotional durability of clothing, please compare also with Burcikova (2017, 2020, 2021). The discussion concludes with a proposition of a *Reflexive framework for sensory fashion and textiles design* (Table 4.1) that opens tangible pathways to implementing sensory analysis of wardrobes into design practice. The approach presented here opens new opportunities especially for the expanding segment of micro and small sustainable fashion businesses (MSEs), who are increasingly recognised as leaders in human-scale, holistic innovations within the fashion sector (Aakko, 2016; Connor-Crabb, 2017; European Commission, 2019).

Methodology

The methodology of the study I conducted between 2015-2018, stemmed from a dialogue between my designer-maker practice and my background in ethnography and cultural studies – a combination that has not been previously used in this area of research. My research role was hence not dissimilar to the notion of 'embodied ethnographer' – someone whose profession gives them entry to an area that is normally not easily accessible to other researchers and whose tacit knowledge from their frequent presence in the researched environment opens new layers for enquiry (Edvardsson and Street, 2007 in Pink, 2015, pp. 20-21). Theory, empirical research, and design practice equally underpinned my enquiry, albeit one or the other may have taken a stronger lead at various stages of the research.

The anthropologist, Robert Murphy (1990) argued that the disadvantage of methods that rely predominantly on participants' verbal statements is that "people often do not do what they *say* they should be doing, or even what they *think* they are doing" (p. 174). Hence, verbal statement data, even if collected from relatively large samples of population, rarely enables access to tacit knowledge or the idiosyncrasies of lived experiences, emotions in particular (Stappers and Sanders, 2004). On the other hand, ethnographies that involve a trusting relationship between the researcher and the researched, utilising a range of complementary methods such as observation, photography, audio/video recordings, notebooks, sketchbooks, or sensory diaries, offer the benefit of comparing verbal accounts to observable behaviour (Murphy, 1990; Pink, 2011, 2012, 2015).

Utilising snowball sampling and word of mouth, starting with the clients of my slow fashion studio, my study included 10 women[1], aged 29-69, interviewed in their homes in several localities in the UK. I chose a narrative approach to interviewing (Elliott, 2005), focusing on what they wanted to tell me, while occasionally steering the conversation with additional questions. One of the key benefits of this approach was that by not asking women to pre-select clothes ahead of the interview, I got to hear not only about successful and 'loved' garments but also about those that failed to satisfy. This enriched my data with material for further cross-analysis.

My method of *wardrobe conversations* (Burcikova, 2020, p. 102) provided data for a multi-level thematic analysis of 450 garments, 20 hours of audio recordings and 2778 photographs, in four iterative stages, including (1) transcription, (2) collating wardrobe narratives and visual documentation of key garments, (3) thematic coding and (4) sketchbook reflections towards final designs, that interpreted each of the key themes through a corresponding garment. This iterative process aligned with Pink's (2015) assertion that analysis in sensory ethnography typically "moves between different registers of engagement with research materials and between different materials" and it also involves "analysing the same materials in different ways" (p. 158).

Pairing selections of my transcribed data with photos from each interview strongly highlighted how visual showing and women's tactile engagement with clothing are inseparable from their verbal descriptions. This corresponds with Pink's (2015 [2009]) observation that our engagement with materials is often quite performative, which means that people tend to "stroke, feel, smell, visually show and as such engage sensorially" with things during interviews (p. 127). Throughout my wardrobe conversations, such sensory interaction with clothing sometimes preceded any verbal descriptions, other times it accompanied women's explanations. Most importantly though, on numerous occasions, touching, stroking, and examining garments in more detail clearly helped women to pin down what may have first seemed hard to put into words.

It is important to note that due to the qualitative approach employed in my study, the findings should not be generalised on the wider population and globally. Accordingly, while I focused on female participants, due to my

---

[1] Julie, Louise, Hanka, Nicola, Golraz, Kathryn, Annabelle, Emma, Mary and Tanya. All participant data have been anonymised, pseudonyms are used throughout.

experience as a designer-maker of womenswear, my findings do not assume a unified female gender identity. The methodology established through this research instead offers some clear methods of analysis and outcomes that can be used in design research and practice and adapted to specific contexts.

As the scope of this chapter does not allow for including the voices of all 10 participants, the narrative here draws on the portrayal method (Christou and King, 2014; Lulle and King, 2015), with focus on the wardrobe of one woman – Louise, corroborating it in later discussion with findings and quotes from across the sample. The wardrobe of Louise was selected for its richness and the insights it provides in relationship to the argument of sensory longevity presented here.

While a wardrobe conversation is a moment in time, a woman's wardrobe is a diachronic process that constantly evolves and can hence never be captured in its entirety. Yet, as the following section will demonstrate, being able to savour at least a fraction of this process through a woman's narrative, still offers a number of valuable clues on the clothes women love to wear and the reasons why they keep wearing the same items over many years.

Louise: I just like it because it's been worn before

Louise's wardrobe

Louise is a French tutor, living in West Yorkshire with her partner and two young children. When I first approached her about my research, Louise felt she did not really have enough clothes that could interest me. Just like several other women I spoke to, she seemed to need reassurance that I was not looking for any extraordinary pieces. She seemed to relax into our conversation when I reiterated that the purpose would be to talk about clothes that she wears on a regular basis.

Louise does not really like new clothes, or clothes shopping as such, because she finds it difficult to find what she would like to wear. This is why her wardrobe mainly consists of clothes handed down to her from other people. Many of these items are up to forty years old. She has several pieces from her grandmother, who had a similar figure, and like Louise, liked simple clothes. Other clothes in Louise's wardrobe come from her mother, her sister and there are also some presents from her best friend. Quite a few pieces, both old and new, are supplied by Louise's mother-in-law. Unlike Louise, she loves clothes shopping and often buys presents for the family in small boutiques in her hometown in northern France.

Louise still also keeps and wears several items from her teenage years in France. For example, the pair of trousers she was wearing at the time of our conversation, is one that she has had for nearly fifteen years (Figure 1).

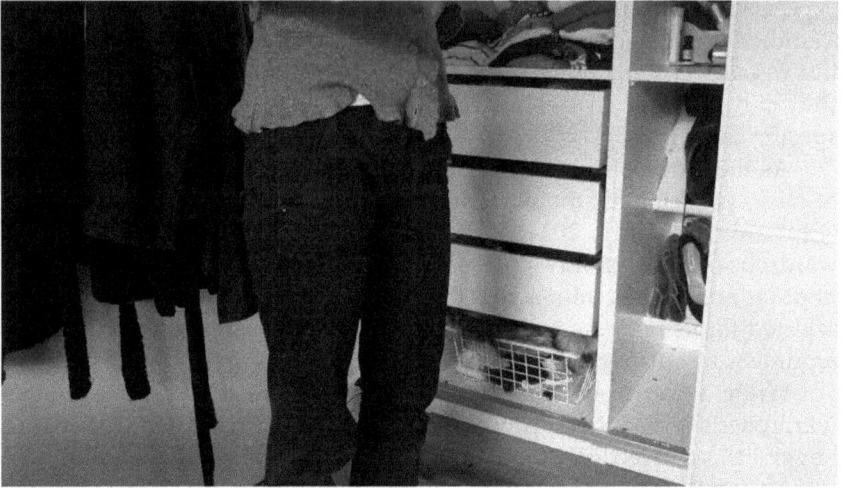

Figure 1. Louise feels extremely comfortable in her rediscovered pair of jeans from her teenage years.

This pair of trousers is the only one I wear. I'm going to be thirty-one and my mum bought them for me when I was about sixteen. I didn't wear them for a long time, but I kept them because I thought they were good quality. Then one day I found them in my mess [laughs] and I was like – "wow"! Now I wear them all the time because they're so comfy! It's a good shape. I've had them for so long and I wear them almost every day.

There is also an even older pair of trousers which she keeps wearing in summer:

I got these pants when I was twelve. Honest! I loved them! I can't wear them outside anymore because they're full of stains and they're probably a bit pinkish. They're supposed to be white. Anyway, my grandma's fixed them several times. She changed the zip, the button, she also had to do some work over here [showing]. And when I grew out of them, she cut the bottom to make them into shorter pants. I still love wearing them in the summer [smiles]. It's like a treasure hunt in my wardrobe.

Another of such treasures is a skirt that Louise remembers her mother wearing when she and her sister were growing up (Figures 2 and 3).

This was just so cool! I used to call it the Pocahontas skirt, when I was little. And then I was old enough, my mum said: do you want it? And I said: yeah – I've always dreamt of having your Pocahontas skirt!

I've worn it a lot, a lot, a lot. Especially when I was pregnant, because it can extend. Unfortunately, that destroyed the skirt – it's stretched. You

Figure 2 and 3. Louise's dream "Pocahontas" skirt that was passed down to her from her mum

can tighten it a little bit but it's lost its glory [laughs]. It's also lost the little pearl that gave it it's Pocahontas style.

Despite the damage and the lost pearl, Louise still keeps the skirt. It has served her well and it may well be useful again sometime in the future – *"you never know"*, she says. Browsing through all the pre-worn clothes in her wardrobe, she is adamant that latest trends or fashion brands are of no interest to her. However, she still has a very clear idea about her own style and the kind of clothes she likes or dislikes. She prefers plain elegance and does not like extravagance in material or design, so she looks for clothes with simple style and straight cut.

Other key considerations for Louise are the softness of material and comfort. Soft touch, she thinks, explains her love of pre-worn clothes because they would have softened through wear and repeated washing. At the same time, she is conscious that quality is a critical factor, *"cause if you wash a Primark top fifteen times then it would be dead, right? – it wouldn't be soft [laughs]"*. The clothes passed onto her by her mother-in-law and other family members have all stood this test of time, just like the bright red work dress that she loves (Figure 4).

This one is also from my mother-in-law. She had it when she was a young girl and she had it adjusted for me. It's very old – you can tell – and it's just really nice. She said: do you want it? I tried it on and it was a bit too wide, so she

Figure 4. A work dress that Louise loves because "it's been worn before".

did that [showing how it's been taken in] – she took it to someone who made it a bit tighter. I love it!

You know, I just like it because it's been worn before. I find that cool [laughs]. Someone loved it and now I love it and maybe someone else will love it one day.

Her all-time favourite, however, is a black day skirt by Sisley, that her mother-in-law used to wear as a student (Figure 5). Louise wears this one all the time, and so when I visited, the skirt was in the dryer. However, she kindly pulled it out to show it to me in the kitchen as I was leaving.

Figure 5. Louise's all-time favourite – a skirt from her mother-in-law's student years.

This is my favourite black skirt. It's easy to wear, it's straight, and I wear it at work with anything. It's very old, right? It's awful – it used to be black and now it's grey. So my favourite clothes are probably not the prettiest [laughs]. Here you see how it used to be really black, but I still love it! It's so versatile! I've been wearing this since I've met my partner – probably good fifteen years now. And look at it! It's still good...

Pre-worn clothing seems to possess the kind of reassuring familiarity and sensory comfort that Louise looks for in her wardrobe. Long history of wear and washing not only softens materials, but it also gradually reshapes each item so that it almost moulds to the body. *You put it on and it's like a second skin really*, she says with a broad smile.

Sensory longevity

The wardrobe portrait of Louise demonstrates the significance of tactile perceptions of clothing for women's clothing relationships. While her wardrobe here serves as an illustration of a case in point, all the women I spoke to repeatedly commented not only on the aesthetic and visual qualities of their clothing, but also, and perhaps more often, on how an item wears on the body or how it makes them feel.

Body plays a vital role in women's decisions about what to wear, Woodward (2007) argues. The tactile perceptions of softness or warmth, as well as the sense of how a garment enables or restricts movement, are all key in such considerations (p. 17). Chong Kwan (2016) further notes that the changing "sensorial materiality" of clothes, as experienced through multiple senses, affects wearers not only physically but also emotionally (p. 284). In addition, while perceptions through multiple senses can often be coordinated and support each other, at other times they can be contradictory and cause confusion (Howes, 2005; Chong Kwan, 2016). The latter often results in mixed feelings about an item of clothing, when for example, a jumper with a beautiful pattern and a lovely colour feels itchy against the skin.

Niinimäki and Armstrong (2013) highlight that it is the physical aspects of garments that we experience first, and while these in themselves do not guarantee either longevity or long-term emotional connection, they are critical to how we respond to a piece of clothing over time (p. 192). Another key insight from my observations of women's sensory engagement with clothing is the significance of apparently small construction details, such as pockets or buttons, for how women relate to the clothes in their wardrobes over time. Pockets, it seems, are humble agents of independence and comfort. Buttons, interestingly, can easily tip the balance between loving or discarding an item of clothing. As I noted earlier, the argument here focuses especially on the relationship of sensory comfort and construction details to clothing longevity, as these have been rarely addressed in discussions to date.

Sensory comfort

In his discussion of Heidegger's *Building, Dwelling, Thinking* (1977), the anthropologist Tim Ingold (2000) remarks that the leading interest of the essay is in determining what it takes "for a house to be a home" (p. 180). My observations of women's sensory engagement with clothing were led by a similar concern. As my explorations progressed, it gradually emerged that research on what lies behind longevity and emotional durability of clothing is in many respects a search for a point when a piece of material worn on the body is no longer perceived as something external to the wearer. If perhaps somewhat counterintuitively, my wardrobe conversations evidence that a strong emotional connection to an item of clothing often manifests itself in that a garment is no longer noticed. Its wearing becomes habitual, and almost unconscious. The feeling of reassuring familiarity makes the experience of wearing it feel like a second nature. In line with Baumann's notion of the home as a place where no defence is required and where there is no need to prove anything (cited in Malicki, 2014, p. 4), such piece of clothing becomes "a home for the body" (Niinimäki, 2010; Chong Kwan, 2016). While such items are then integral to how we feel and how we experience the world around us, their success relies on the fact that their wearing is effortless and almost unnoticed by the wearer. This is evidenced in my interviewees' multiple references to garments that "feel like not wearing anything" (for example by Hanka, Julie, Louise, and Mary). All such instances are strongly linked to bodily comfort negotiated by the softness of materials, often enhanced through long-term use.

Several materials have been recognised in wardrobe research for their capacity to improve and become more comfortable through wear. For example, wool and leather are frequently mentioned as materials that age "gracefully", and so they often feature in narratives of favourite items and long-term use (see e.g. Woodward, 2007; Niinimäki, 2010, 2014; Fletcher, 2016; Connor-Crabb, 2017; Ravnløkke, 2019). Denim is an especially compelling point in case. Its popularity and wide appeal is often linked to its inherent quality of moulding to the body and aging with the wearer (Candy, 2005; Woodward, 2007; Miller and Woodward, 2011).

In sum, soft and light garments seem to possess the crucial capacity to be appreciated for being "unnoticed" by wearers. Just like health is the blessing of being unaware of one's body (Murphy, 1990), comfort experienced through the soft feel and the lightness of fabric seems to be a state of being unaware of one's clothes. Crucially, in terms of the argument presented here, both these qualities also tend to improve with time, through continuous wear and laundering. Tanya's beloved blue dress that she has worn for over fifteen years has softened to an extent that it is now "incredibly comfortable to wear". Importantly for her lifestyle of frequent travel, it's worn quality also means that it now folds very easily into her suitcase. As highlighted in Louise's wardrobe portrait above, the unique worn quality that a clothing acquires over time is also one of the main reasons for her preference of old clothes over new items.

Details

In addition, the evidence from my wardrobe conversations provides further support for Fletcher's (2016) suggestion that "our search for satisfaction – so often the motivation behind a new round of consumption of whole fashion pieces – is channelled through uncovering and noticing the details" (p. 283).

As I already noted in the methodology section, my focus on photographing the ways in which women handled the clothes they chose to show me, led me to noticing how their hands followed their favourite details. Emma, for example, pointed out the fabric covered buttons that were her favourite feature of her mum's old dress. Tanya proudly showed me the inside of the pocket on her beloved travel dress – a cherished reminder of its now washed-out original colour. Louise's favourite "Pocahontas skirt" (see Figures 2-3), named after a plastic pearl at the end of the drawstring, is another great illustration here. Subtle and often hidden construction details can capture the intimacy and the richness of meaning only known to the wearer and so they can be vital for a garment's long-term appreciation.

Buttons and other decorative details

As the following extracts illustrate, buttons, belts, and decorative details such as applique or embroidery were frequently mentioned in connection to favourite items:

> *I like the fact that you've got these sort of brocade things, the velvet and then it's done it also on the pocket (Kathryn)*

> *This is very old…I love it because of these flowers (Golraz)*

> *I liked it because of the belt (Mary)*

> *I like things like this – I like pearl buttons (Nicola)*

> *It's a feature on the sleeves – I think that's quite nice – it gives a little something to the jacket (Louise)*

Kathryn and Emma especially emphasized their love of buttons. They both seem to have jars of vintage buttons which they occasionally use to alter and liven up some items. *"I quite often change buttons on things,"* Emma says, *"because it makes quite a difference, doesn't it?"*

Buttons, my research confirms, can decide whether a piece of clothing finds its use within its owner's wardrobe. This strongly corresponds with Fletcher's (2016) point that, "many times the stories from the public suggest that it is the details and components of garments that hold the key to satisfying use" (p. 238). Hanka offers a good illustration here, as she tells me that buttons were a great disappointment to her when she first received a dress from one of her favourite designers bought on E-bay. Similarly, Annabelle admits that she *"felt a bit rotten"* for throwing away a jacket from her mother's two-piece, because *"the buttons were just too big and looked dated"*.

In addition, garment details are often also seen as important indicators of quality. This comes through especially clearly as Golraz explains that to assess the quality of a garment when clothes shopping, she first looks at details. She shows me the zip on one of her favourite day dresses, gently unzipping and then closing it again. She repeats this several times as we speak, gently savouring the process. Her love of this dress is strongly linked to the way the zip has been inserted:

> *Because when I wear it, I kind of think that I am taking care of myself – because somebody took care of this. This has been loved, you know [smiles]. That's what I like.*

Similarly, Julie points out the lining and the mismatched buttons on a designer jacket that she had bought on a shopping trip with a friend. While she is adamant that she does not care about designer brands per se, she admits that *"these kind of details"* that she likes, are more likely to come with the more expensive designer pieces. Like Golraz, Julie too says that attention to detail makes her feel that an item she is wearing was well made. It is this sign of quality, rather than a brand, that she looks for and appreciates in a piece of clothing.

Pockets

It doesn't have pockets, that's its only fault, Hanka tells me as she hands me a dress that she loves wearing in summer. Pockets, as it later turns out, are key to the success of any item in her wardrobe. Mary and Emma do not especially look for pockets when clothes shopping, but along with the rest of my interviewees they too agree that pockets are an appreciated and extremely useful feature in clothing. While Mary feels that it is also important for pockets to work with a garment's cut, she admits that they have their place in most garments, work jackets especially:

> *It's nice to have pockets. It's gonna be irritating on the odd occasion when you're wearing something that doesn't have pockets. It is irritating not having pockets on a jacket (Mary)*

As Summers (2016) argues, having pockets to put things into gives women independence and a freedom to walk around unburdened by extra items such as purses. This comes through especially strongly in Kathryn's description of one of her longest standing pieces of clothing – a denim jacket that she associates with holidays. She loves it especially because she can put her keys and anything else she needs in her pockets when she wears it with a dress in summer. The same is reiterated by Golraz, who says she always has something to put in her pockets. If her clothes do not have them, *"it's a problem because I have to put things in my bag then"*.

In sum, pockets seem to be a strong feature of many favourite garments. If my interviewees liked an item without pockets, their absence was often commented on, just like in Hanka's above description of her pocketless dress. The lack of pockets seems to be *"a shame"* and *"a disadvantage"* of a piece of clothing.

Design for sensory longevity

The discussion above demonstrates that women's tactile perceptions of clothes can significantly affect how they feel about individual items in their wardrobes. The success of favourite pieces often manifests itself in that their wearing is effortless and almost unnoticed by their wearers. The feeling of comfort, often negotiated trough the light weight and soft touch of materials, can thus be a critical factor in longevity and emotional durability of clothing.

Another key finding from my wardrobe conversations relates to the significance of construction details for satisfying use. Surprisingly, aside from few exceptions (WRAP, 2013; Fletcher, 2016), these have received little attention in research on fashion and sustainability more broadly, and design for longevity and emotional durability more specifically. More questions on garment details therefore need to be raised in the future by researchers and design practitioners with interest in clothing longevity.

While noticing the significance of sensory perceptions in everyday experiences with clothing, it seems striking that sensory aspects of design hardly feature in design education – also highlighted by Sonneveld (2004). In her *Doing Sensory Ethnography,* Pink (2015 [2009]) recommends that sensory ethnographers prepare for their fieldwork by an auto-ethnographic exercise which can help them develop an understanding of their own sensory perceptions of the world (p. 60). In light of the empirical evidence presented here, I propose that designers who wish to design clothes that can have long-term relevance in people's wardrobes, would usefully benefit from a similar auto-ethnographic exercise.

Sissons (2016) argues that designers can hardly expect people to want to wear their creations if they themselves would not want to wear them. Echoing her approach, I propose that if designers develop a deeper understanding of their own sensory responses to the clothes they wear, they can more easily implement multi-sensory considerations into their creative work. As discussed throughout this chapter, the evidence from my wardrobe conversations strongly suggests that this could considerably improve everyday experiences of people who will go about their lives wearing their designs.

Reflexive framework for sensory fashion and textiles design

The reflexive framework in the table below (Table 1) was developed as an extension of the sensory wardrobe methodology established through this research. Its purpose is to help cultivate designers' sensitivity towards multi-sensory perceptions of clothing and to focus imagination beyond visual aspects of design. To date, it has been utilised in the application of this research to designer-maker practice (see Burcikova, 2020 for more detail). It has also been used as a teaching tool at London College of Fashion, University of the Arts London and the School of Arts, Design and Architecture at Aalto University. The framework can be equally adapted for use within design teams in fashion and textiles MSEs and larger organisations, to facilitate contextual understanding of clothing longevity and to shift creative focus away from the still prevailing cult of newness.

Table 1. Reflexive framework for sensory fashion and textiles design

To better understand how sensory perceptions influence our relationships with the materials and clothes we wear, the best place to start is our own wardrobe. How can your feelings about your own clothes help you imagine how your own designs will feel when other people wear them?	
Day 1	Look through your wardrobe and pick three items that you like wearing. Focus on those clothes that you will be able to wear as often as possible over the next couple of weeks. During this time, start a diary where you will record every detail that comes into your mind while wearing each of these items. You can use text, images, sound or/and video recordings, material swatches or any other ways that help you collect and document your experiences, feelings, and impressions.
Week 1-2	Focus especially on the below, but explore and record much more*: • Why do you like this piece of clothing? • Why do you put it on in the morning? • How does it make you feel? • Is the material suitable for the things you will be doing? How so? • Is the style suitable for the things you will be doing? How so? • Do you need to get changed later in the day? If so, why? • Can you name the material? Do you know where it comes from? • How does the material feel on the body - do you feel warm, cold, is the material pleasant against your skin? • How many different materials are used in the garment – observe also components such as buttons, linings, pockets, drawstrings, zips. How do each of these feel (warm, cold, soft, comfortable, uncomfortable)? • What kind of sounds do they make? • How many colours are there? If there are more than one, do you like one more than others? Alternatively, is it the colour combination that you find attractive? Why so?

	• Does the material crease? Can you wash it? Can you iron it? • How many times did you need to do this over the two weeks? • Is the item new? How and why did you get it? • Is it old? Do you remember how long you have had it? • Does it need any repairs? • If so, would you be able to do these yourself? • What do you like the most about this item? • What would you improve about the material or the style? *Resources for inspiration: Burcikova (2017, 2020, 2021), Sampson (2020), Stasiulyte (2018, 2020), Fidler-Wieruszewska (2020).
Day 14	At the end of the two-week period, prepare a 2-5 minute presentation, a video, or sketchbook sequence that will capture these experiences in a way that will help you implement them in your future designs.

Conclusion

The empirical evidence presented in this chapter alongside strategies for incorporating sensory analysis of wardrobes into design, will be of interest to a breadth of researchers and students of fashion concerned with clothing longevity and sustainable fashion futures. Critically though, this way of working opens new pathways for designers who wish to cultivate "emotional and sensorial closeness to their users" (Von Busch, 2018). It is therefore especially suitable for the expanding segment of micro and small sustainable fashion businesses (MSEs), who are increasingly recognised as leaders in human-scale, holistic innovations within the fashion sector (Aakko, 2016; Connor-Crabb, 2017; European Commission, 2019). While the mainstream narrative of fashion is still ruled by large players and high-street chains, the global fashion sector largely comprises of micro and small businesses (FSP, 2018-21; Statista, 2017; European Commission, 2019).

In this sense, the approach to fashion design and making that this chapter introduces sits outside the production methods used in the "current condition" (Fletcher, 2016, p. 272) of the fashion industry at large. Focus on sensory longevity of clothing is not an approach to be scaled up, thus residing within the current narrative of growth and scale. Rather, it is a way of working to be "scaled across" (Connor-Crabb, 2017) by those whose operations, like my own designer-maker practice, aim to counteract the mindset of clothing disposability by embracing models that enable rich customer engagement and iterative feedback on design (Williams et al., 2021). Larger scale design and production industries of fashion and clothing can use this research as a point of reference to reconceptualise their practices towards long-term customer satisfaction, shifting away from an outdated and wasteful model that continues to accelerate the climate crisis.

References

Aakko, M. (2016). *Fashion in-between: Artisanal Design and Production of Fashion* (PhD thesis). Helsinki: Aalto University.

Banim, M. and Guy, A. (2001). Dis/continued selves: Why do women keep clothes they no longer wear? *In:* Guy, A., Green, E. and Banim, M. (Eds.), *Through the Wardrobe: Women's Relationships with their Clothes.* pp. 203-219. Oxford: Berg.

Bridgens, B. and Lilley, D. (2017). Understanding material change: Design for appropriate product lifetimes. *In:* Bakker, C. and Mugge, R. (Eds.), *Product Lifetimes and the Environment (PLATE) 2017 – Conference Proceedings.* pp. 54-59. Delft, Netherlands: Delft University of Technology and IOS Press.

Burcikova, M. (2017). Satisfaction matters. Design that learns from users' sensory and emotional responses to clothing. *In:* Bakker, C. and Mugge, R. (Eds.), *Product Lifetimes and the Environment (PLATE) 2017 – Conference Proceedings.* pp. 60-64. Delft, Netherlands: Delft University of Technology and IOS Press. Available at: https://ebooks.iospress.nl/volumearticle/47843 (Accessed: 15 June 2022).

Burcikova, M. (2020). *Mundane Fashion: Women, Clothes and Emotional Durability* (Unpublished doctoral thesis). University of Huddersfield: Huddersfield.

Burcikova, M. (2021). Mundane durability: The everyday practice of allowing clothes to Last. *PLATE: Product Lifetimes and the Environment*, May 26-28, Limerick [on-line]. Available at: https://ulir.ul.ie/bitstream/handle/10344/10201/Burcikova_2021_Mundane%20durability%20The%20everyday.pdf (Accessed: 15 June 2022).

Candy, F. (2005). The fabric of society: An investigation of the emotional and sensory experience of wearing denim clothing. *Sociological Research Online,* 10(1): 124-140.

Chapman, J. (2009). Design for (emotional) durability. *Design Issues,* 25(4): 29-35.

Chapman, J. (2015 [2005]). *Emotionally Durable Design: Objects, Experiences and Empathy.* London: Routledge.

Chong Kwan, S. (2016). *Making Sense of Everyday Dress: Integrating Multi-sensory Experience within our Understanding of Contemporary Dress in the UK* (Unpublished doctoral thesis). London: University of the Arts.

Christou, A. and King, R. (2014). *Counter-diaspora: The Greek Second Generation Returns 'Home'.* Cambridge MA: Harvard University Press.

Connor-Crabb, A. (2017). *Fashion Design for Longevity: Design Strategies and their Implementation in Practice* (Unpublished doctoral thesis). Brighton: University of Brighton.

Cooper, T. (2010). The significance of product longevity. *In:* Cooper, T. (Ed.), *Longer Lasting Products: Alternatives to the Throwaway Society.* pp. 3-36. Farnham: Gower.

Cupchik, G.C. (1999). Emotion and industrial design: Reconciling meanings and feelings. *In:* Overbeeke, C.J. and Hekkert, P. (Eds.), *Proceedings of the 1st International Conference on Design and Emotion*. pp. 75-82. Delft: Delft University of Technology.

Edvardsson, D. and Street, A. (2007). Sense or no-sense: The nurse as embodied ethnographer. *International Journal of Nursing Practice*, 13: 24-32.

Elliott, J. (2005). *Using Narrative in Social Research: Qualitative and Quantitative Approaches*. London: SAGE.

European Commission (2019). *Support Report Mapping Sustainable Fashion Opportunities for SMEs*. Available at: https://ec.europa.eu/growth/sectors/fashion/fashion-and-high-end-industries/fashion-and-high-end-industries-eu_en (Accessed: 15 June 2022).

Fidler-Wieruszewska, A. (2020). *Yellow Jacket Exploration by Poly Cotton*. April 28 [online]. Available at: https://www.youtube.com/watch?v=YiVGT9hDMoA (Accessed: 15 June 2022).

Fletcher, K. (2014 [2008]). *Sustainable Fashion & Textiles: Design Journeys*. London: Routledge.

Fletcher, K. (2016). *Craft of Use: Post-growth Fashion*. Abingdon: Routledge.

FSP Rethinking Fashion Design Entrepreneurship: Fostering Sustainable Practices (2018-21) [online]. Available at: https://www.sustainable-fashion.com/fsp (Accessed: 15 June 2022).

Gimeno Martinez, J.C., Maldini, I., Daanen, H.A.M. and Stappers, P.J. (2019). Assessing the impact of design strategies on clothing lifetimes, usage and volumes: The case of product personalisation. *Journal of Cleaner Production*, 210: 1414-1424.

Gnanapragasam, A., Cooper, T., Cole, C. and Oguchi, M. (2017). Consumer perspectives on product lifetimes: A national study of lifetime satisfaction and purchasing factors. *In:* Bakker, C. and Mugge, R. (Eds.), *Product Lifetimes and the Environment (PLATE) 2017 – Conference Proceedings*. pp. 144-148. Delft, Netherlands: Delft University of Technology and IOS Press.

Haines-Gadd, M., Chapman, J., Lloyd, P., Mason, J. and Aliakseyeu, D. (2017). *Emotionally Durable Design Framework*. Brighton: Philips Lighting Research and University of Brighton.

Heti, S., Julavits, H. and Shapton, L. (2014). *Women in Clothes: Why We Wear What We Wear*. London: Particular Books.

Howes, D. (Ed.) (2005). *Empire of the Senses. The Sensual Culture Reader*. Oxford: Berg.

Ingold, T. (2000). *The Perception of the Environment: Essays on Livelihood, Dwelling and Skill*. London: Routledge.

Jordan, P.W. (2000). *Designing Pleasurable Products: An Introduction to the New Human Factors*. London: Taylor & Francis.

Laitala, K., Boks, C. and Klepp, I.G. (2015). Making clothing last: A design approach for reducing the environmental impacts. *International Journal of Design*, 9(2): 93-107.

Lulle, A. and King, R. (2015). Ageing well: The time-spaces of possibility for older female Latvian migrants in the UK. *Social & Cultural Geography*, 17(3): 444-462.

Malicki, H.E.O. (2014). *Home-work: A Study of Home at the Threshold of Autoethnography and Art Practice* (Unpublished doctoral thesis). Edinburgh: The University of Edinburgh.

Miller, D. and Woodward, S. (2011). *Global Denim*. Oxford: Berg.

Mugge, R., Schoormans, J.P.L. and Schifferstein, H.N.J. (2005). Design strategies to postpone consumers' product replacement: The value of a strong person-product relationship. *The Design Journal*, 8 (2): 38-48.

Mugge, R. (2008). *Emotional Bonding with Products: Investigating Product Attachment from a Design Perspective*. Saarbrucken: VDM Verlag.

Murphy, R. (1990). *The Body Silent*. New York: W.W. Norton.

Niinimäki, K. (2010). Forming sustainable attachments to clothes. *In:* Proceedings of the 7th International Conference on Design & Emotion, IIT, Chicago, Oct 4-7.

Niinimäki, K. and Hassi, L. (2011). Emerging design strategies in sustainable production and consumption of textiles and clothing. *Journal of Cleaner Production*, 19(2011): 1876-1883.

Niinimäki, K. and Koskinen, I. (2011). I love this dress, it makes me feel beautiful! Empathic knowledge in sustainable design. *The Design Journal*, 14(2): 165-186.

Niinimäki, K. and Armstrong, C. (2013). From pleasure in use to preservation of meaningful memories: A closer look at the sustainability of clothing via longevity and attachment. *International Journal of Fashion Design, Technology and Education*, 6(3): 190-199.

Niinimäki, K. (2014). Sustainable consumer satisfaction in the context of clothing. *In:* Vezzolli, C. Kohtala, C. and Srinivasan, A. (Eds.), *Product-Service System Design for Sustainability*. pp. 218-237. Sheffield: Greenleaf Publishing.

Norman, D.A. (2004). *Emotional Design: Why We Love (or Hate) Everyday Things*. New York: Basic Books.

Pink, S. (2011). Ethnography of the invisible: how to 'see' domestic and human energy. *Ethnologia Europaea: Journal of European Ethnology*, 41(1): 117-128.

Pink, S. (2012). *Situating Everyday Life: Practices and Places*. London: SAGE.

Pink, S. (2015 [2009]). *Doing Sensory Ethnography*. London: SAGE.

Ravnløkke, L. (2019). Design af strikbluser til lang levetid: Strikkede prototype som redskab for brugerdialog I designprocessen' [Design of knitted jumpers for longevity: Knitted prototypes as a tool for user dialogue in the design process] (Unpublished doctoral thesis). Kolding: Design School Kolding.

Roberts, P. (2015). *The Impulse Society: What's Wrong with Getting What We Want*. London: Bloomsbury.

Sampson, E. (2020). *Worn: Footwear, Attachment and Affects of Wear*. London: Bloomsbury.

Shove, E., Trentmann, F. and Wilk, R. (2009). Time, Consumption and Everyday Life. Oxford: Berg.

Sissons, J. (2016). Panel discussion at The Second International Conference for Creative Pattern Cutting – Creative Cut, University of Huddersfield, 24-25 February.

Skjold, E. (2014). The Daily Selection (Unpublished doctoral thesis). Kolding and Copenhagen: Design School Kolding and Copenhagen Business School.

Skjold, E. (2016). Biographical wardrobes – A temporal view on dress practice. *Fashion Practice: The Journal of Design, Creative Process and the Fashion Industry*, 8(1): 135-148.

Sonneveld, M. (2004). Dreamy hands: Exploring tactile aesthetics in design. *In:* McDonagh, D. , Hekkert, P., Van Erp, J. and Gyi, D. (Eds.), *Design and Emotion:*

The Experience of Everyday Things. pp. 260-265. London: Taylor & Francis Publishers.

Stappers, P.J. and Sanders, E.B.N. (2004). Generative tools for context mapping: Tuning the tools. *In:* McDonagh, D., Hekkert, P., Van Erp, J. and Gyi, D. (Eds.), *Design and Emotion: The Experience of Everyday Things.* pp. 85-89. London: Taylor & Francis.

Stasiulyte, V. (2018). *Sonic Fashion Library* [online] Available at: http://sonicfashion. se/ (Accessed: 15 June 2022).

Stasiulyte, V. (2020). *Sound to Wear.* September 24 [online]. Available at: https:// www.youtube.com/watch?v=YU4KCpEPavA(Accessed: 15 June 2022).

Statista (2017). *Apparel and Clothing Market Europe – Statistics and Facts* [online]. Available at: https://www.statista.com/topics/3423/clothing-and-apparel-market-in-europe/ (Accessed: 15 June 2022).

Summers, Ch. G. (2016, Sep 19). The politics of pockets: The history of pockets isn't just sexist, it's political. *Vox,* 19 September. Available at: https://www. vox.com/2016/9/19/12865560/politics-of-pockets-suffragettes-women (Accessed: 15 June 2022).

Van Hinte, E. (1997). *Eternally Yours: Visions of Product Endurance.* Rotterdam: 101 Publishers.

Von Busch, O. (2018). *Vital Vogue. A Biosocial Perspective on Fashion.* New York: SelfPassage.

Walker, S. (1995). *Sustainable by Design: Explorations in Theory and Practice.* London: Earthscan.

Walker, S. (2006). Object lessons: Enduring artifacts and sustainable solutions. *Design Issues,* 22(1): 20-31.

Williams, D., Burcikova, M. and Buchan-Ng, M. (2021). *Fashion as Sustainability in Action: A Guide for Fostering Sustainable Prosperity in Micro and Small Fashion Businesses.* UAL: Centre for Sustainable Fashion. Available at: https://www. sustainable-fashion.com/fsp (Accessed: 15 June 2022).

Woodward, S. (2007). *Why Women Wear What They Wear.* Oxford: Berg.

WRAP (2013). *Design for Longevity: Guidance on Increasing the Active Life of Clothing.* Available at: https://www.academia.edu/31425358/Cooper_T_Hill_H_ Kininmonth_J_Townsend_K_and_Hughes_M_2013_Design_for_Longevity_ Guidance_on_increasing_the_active_life_of_clothing_Report_for_WRAP_ Waste_and_Resources_Action_Programme_Banbury (Accessed: 15 June 2022).

WRAP (2015). *Clothing Durability Report. Banbury: Anthesis & Waste and Resources* Action Programme.

PART II
Use

European Circular Economy Perspectives on Fashion and Textile Consumer Behaviour

Han, S.L.C.*, Blanco-Velo, J., Boiten, V.J. and Tyler, D.

Nottingham Trent University, UK
e-mail: *sara.han@ntu.ac.uk

Introduction

The sustainable consumption of clothing spans several stages, from the decisions made at the level of purchasing, to the various phases of use, care, reuse, and disposal. Consumer behaviour and decisions on clothing purchases have been studied from a variety of angles, such as social psychology, consumer research, fashion marketing and identity studies. Less attention has been devoted to disposal behaviour (Domina and Koch, 1999; Gracey and Moon, 2012). Studies covering the disposal phase have often looked into the eventual destination of used clothing and textiles, but less so into the perceptions, reasons and convictions that drive the decision-making on how and where to discard a used garment (Domina and Koch, 1999; Gracey and Moon, 2012; Laitala, 2014).

Reuse and recycling are an important pathway towards reducing the climate and environmental impacts of textile production (Dahlbo et al., 2017). Across the globe, it is considered that the equivalent of one garbage truck of textiles is landfilled or incinerated every second, while just 1% of clothing materials is recycled into new clothing (Ellen MacArthur Foundation, 2017). In a more circular system however, the value of textile resources is maximised post-usage: with a shift from product back into material – fibres and other materials receiving customised treatment for reuse and recycling. The policy framework for a "circular" system of production and consumption is being developed and trialled today, as part of the efforts to meet the Paris Agreement climate targets and to implement the Sustainable Development Goals (SDGs). As an example, the European revised legislative framework on waste, which entered into force in 2018, stipulates a common EU target for recycling 65% of municipal waste by 2035, and requires separate collections for textiles to be set up by 2025. But the apparel sector is equally moving ahead: at the August 2019 G7 Summit, a coalition of global fashion and textile companies signed the Fashion Pact, committing to zero greenhouse gas emissions by 2050, restoring natural ecosystems and biodiversity and removing single-use plastics.

Consumers are a crucial element in the transition to more sustainable models of production and consumption. Changing business models and bringing new products and services to the market, requires equally a change in mindset among consumers. To accelerate the awareness-raising and widespread adoption of more circular practices, the European Economic and Social Committee (EESC) calls for a strategic shift towards education, lifelong learning, and information. Consumers need to be enabled, according to the EESC, to make active choices based on accurate information detailing products' social and environmental footprint. This involves information on emissions, biodiversity conservation, estimated lifespan, possibility of obtaining spare parts, options for repair, etc.

Such roles for consumers as active decision-makers and agents of sustainable consumption raises questions around consumers' current levels of information and awareness, as well as around their preferences towards receiving such information. Evidently, concepts such as knowledge and awareness are complex parameters to measure and assess, with the "perception-behaviour link" particularly challenging to unravel (Chartrand, 2005).

This chapter therefore commences with a focus on exposing consumers' behavioural patterns when it comes to the phases of purchase, use and disposal. Perceptions of disposal and the convictions, ideas and arguments that drive this behaviour are evaluated, e.g., why a charity shop is chosen over kerbside collection or vice versa. The range of perceptions that consumers hold and develop, are shaped by media, word of mouth, education, advertisements, reputation, past experiences and so forth. The second phase of this chapter seeks to identify the sources of information that consumers rely on to inform their behaviour and their preferred information channels.

Background

The barriers to behaviour change

The Theory of Planned Behaviour, as shown in Figure 1 (Ajzen, 1985, 1987) suggests that attitudes toward the behaviour, subjective norm, and perceived behavioural control – all impact intentions, which can then predict behaviour in a specific context. Ajzen (1991) explains that an attitude towards a behaviour is the extent to which the behaviour being considered is likely to be favourable or unfavourable to the consumer. The subjective norm refers to the extent to which the consumer perceives any social pressure to carry out or not carry out the behaviour and control refers to the extent to which the consumer perceives the ease of carrying out the behaviour.

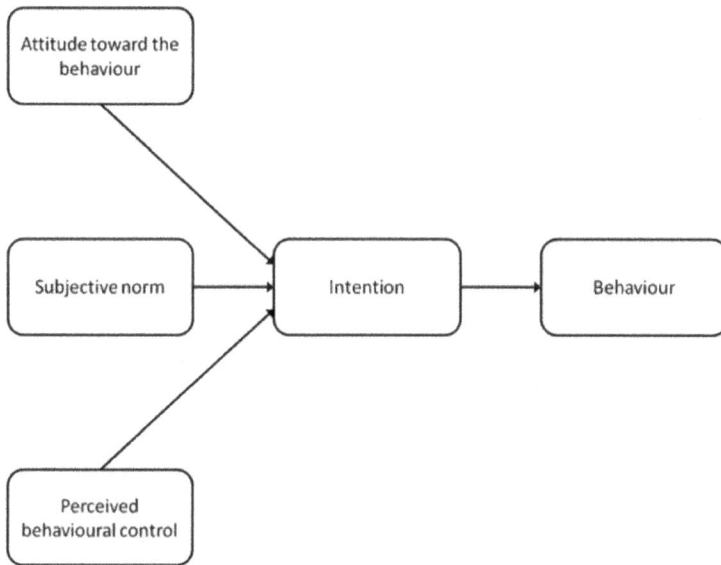

Figure 1. The theory of planned behaviour (Ajzen, 1991)

While it is widely accepted that some form of communication is critical to the development of consumer action and changes in consumption behaviours, the research findings of academicians such as Hobson (2001, 2003), Jackson and Michaelis (2003) and Jackson (2004), propose that an information-led approach to behaviour change misses a great deal of what actually shapes consumption activities. Sector level and academic research within the context of UK fashion and clothing confirms the suggestion that the provision of information to drive change in consumer behaviour is hampered by a lack of consumer understanding (Thomas, 2008; Beard, 2008; Fisher, 2008; Harris et al., 2016; Reimers et al., 2016). Despite ongoing research, why this is the case remains unclear.

Hiller Connell (2010) provides in-depth analysis of the literature which explores the barriers to behaviour change in relation to the purchase and consumption of textiles and clothing. These are explained through the concepts of **internal** and **external** barriers. In evaluating the work of Hiller Connell, Harris et al. (2016, p. 311) summarise **external barriers** as follows:

- Limited availability of sustainable clothing outlets
- Restricted styles
- Lack of desired size and fit
- Lack of resources to accommodate the price of products
- Poor presentation of second-hand clothing
- Social expectations regarding the conventions of professional dress

and **internal barriers** as:

- Lack of concern for environmental issues
- Limited knowledge about the environmental impacts of clothing
- Negative attitudes towards sustainable clothing
- Demographic characteristics
- Social and cultural norms
- Motivation
- Locus of control
- Perceived time and effort

In evaluating this list, desirability, price and social norms have an impact upon decision making, concepts that are supported by the findings of Tomolillo and Shaw (2003), Bray (2011) and Connell (2010). Literary findings are consistent in the expression of consumers' unwillingness to pay a premium for products that support sustainable beliefs (McEachern et al., 2010; Sudbury and Boltner, 2011) and not necessarily due to affordability. Ritch (2015) recognises that in considering a sustainable garment to a non-sustainable garment, the assumption is that it will indeed cost more. Several authors have assessed the willingness of consumers to pay a premium. Findings suggest that, while on average, people are willing to consider a 25% premium for products made with organic cotton (Hustvedt and Bernard, 2008; Ellis et al., 2012), the final decision to purchase continues to be based upon conventional product attributes of price, style and brand rather than a products' inherent sustainable qualities (Koszewska, 2013; Reimers et al., 2016). The findings of McNeill and Moore (2015) indicate that the pull of fast fashion cannot be ignored as a factor in what appears to have become a 'social conditioning' to seek high fashion low-cost products, despite evidence of a willingness to pay a premium for organic or sustainably produced food.

Ritch (2015) provides perhaps the most salient findings in terms of leading further discussion of how and why consumers are hampered in transferring their understanding of, for example, sustainably produced food to clothing. In concert with the findings of Sisco and Morris (2012), Ritch proposes that in order to perceive the benefits of sustainable or environmentally beneficial behaviours in relation to clothing, they must first have an in-depth knowledge of the concepts that inform its production use, and disposal. Discussion suggests that in order to develop attitudes that will convert to behaviour in the purchase, appropriate use and sustainable disposal of clothing, then consumers must be 'product literate'. Product literacy is a concept which, to date, has not been fully considered in the literature regarding the attitude knowledge or perception behaviour gap in relation to sustainable clothing; nor has it been considered as an antecedent to consumer knowledge and a means to reducing uncertainty in the evaluation of product or disposal related information.

Product literacy

Product literacy is the degree to which consumers have the capacity to locate, obtain, evaluate, apply and communicate basic information needed to make appropriate product related decisions (Kopp, 2012). Research suggests that product literacy is contextual. The development of product literacy depends upon the product, on the information provided about the product and upon a broad range of individual consumer characteristics. Pappalardo (2012) suggests that a person attains product literacy when he or she possesses the tools necessary to determine if a particular product or service will meet his or her goals given his or her limited resources including limited wealth, limited time, and limited household production capabilities.

Kopp (2012) proposes that the development of product literacy is contingent upon three conditions. First, consumers need to estimate the net benefit of obtaining product information. Consumers need a sense of the likely costs and benefits of obtaining additional information and knowledge. Second, consumers need to comprehend this information. Third, consumers need a way to evaluate this information and relevant choices.

Product literacy, according to Kopp (2012) is a key ability in a consumption-driven society, composed of a fundamental set of skills and knowledge, needed to make "satisfying" individual purchase choices but also to influence general health, economy and societal wellbeing. Kopp refers to regulators who either presume to "educate" consumers by providing "more information" or attempt to simplify or frame the existing information to a level that consumers can understand. Kopp highlights that product literacy does not require that consumers have perfect information, only that they are able to access enough information to make a reasonably good decision.

Consumer knowledge

Consumer knowledge theoretically consists of two dimensions: familiarity and product knowledge (Johnson and Russo, 1984; Kang et al., 2013). Product knowledge can be summarised as the body of product class information that is held in the memory of a consumer. Familiarity refers to the accumulated consumption experiences (Johnson and Russo, 1984) and is considered critical in the development of the perceived personal relevance of a product in terms of her personal lifestyle, values and self-image (Kang et al., 2013). Perceived personal relevance is considered critical to the development of product involvement (Zaichkowsky, 1985).

Park et al. (1994) suggests that there are two forms of consumer knowledge. These are objective knowledge: accurate information about the product class stored in long-term memory and self-assessed knowledge or subjective knowledge: people's perceptions of what or how much they know about a product class. The work of Ellen (1994) suggests that consumers with higher degrees of familiarity or product knowledge were more likely to use intrinsic (i.e., physical product) cues to assess product alternatives while those with lesser knowledge would typically rely upon extrinsic cues (i.e., attributes not related to the physical product, i.e., price). Ellen (1994) suggests that each form of consumer knowledge has a different effect on behaviour. While one consumer may have the requisite knowledge to make informed decisions, he or she may still not feel well informed because the "right" choice is not perfectly clear in all situations giving rise to uncertainty about the product. The author acknowledges the complexity caused by a lack of "hard" meanings for the phrases used to communicate sustainable and ecological benefits (i.e., recyclable, recycled, degradable etc.) proposing that consumers with lower levels of knowledge may find it difficult to make "good" choices because of the potential for confusion.

Involvement

Zaichkowsky (1985, p. 342) defines involvement as a person's perceived relevance of the object (message or product) based on inherent needs, values, and interests. That is, the higher the degree of relevance of a message or a product to a consumer, the higher that consumer's level of involvement with the information or, in the case of this research, the sustainable fashion product (Josiam et al., 2005). Foxall et al. (1998) recognised involvement for the role it plays in attitude formation. Under high involvement conditions, consumers engage in an extended problem-solving process (Zaichkowsky, 1985). Research to date suggests that the pro-environmental or sustainably committed consumer can be defined as having high involvement with sustainable fashion products where the mainstream consumer's involvement is low.

High involvement implies greater relevance to the self (O'Cass, 2000) and has the potential to lead to enduring involvement, which is stable over time. Thus the higher the level of involvement, the more likely a consumer is to seek out information with which to evaluate possible alternatives. This outcome is less likely in low involvement as products are considered as having little relevance to consumers or, possibly, where the consumer is less product literate (Kopp, 2012), the information about the product is not understood. Michaelidou and Dibb (2006) cite the Laaksonen (1977, p. 445) definition of response involvement as a behavioural process and thus a "means to mediate information search". In his extensive review of the literature related to fashion involvement, Naderi (2013) provides evidence that, to date, the consequences of fashion involvement studied in the articles reviewed are either behavioural (i.e., search behaviour) or attitudinal (i.e., attitude durability) constructs. Naderi (2013, p. 101) calls for consideration of the antecedents and consequences of product involvement: i.e., information processing and its impact upon decision-making factors.

Antecedents to involvement

In considering the details of the variables that precede involvement, these may be summarised in relation to the extant literature in terms of three inherent factors; the characteristics of the person (consumer) (Zhaikowsky, 1985), the physical characteristics of the stimulus (the form of media text, accessibility of media message or the sustainable fashion product) (Scheufele's, 1999) and finally, grounded in the perspective of Houston and Rothschild (1977), variations in the situation of the person/consumer. In terms of physical characteristics, demographic characteristics such as age and gender are recognised to influence personal motivations (Kim and Hong, 2011).

What is not considered in this model of behaviour are the antecedents to attitude formation, the situations that influence the subjective norm or the situations that influence the perceptions of control. Within the context of the present study, the concept of the values-action gap would appear to provide some insight to the nature of the antecedents to positive attitude formation. Markkula and Moisander (2012), in considering the limits to the positive effects of policy development, draw attention to a disquieting values-action gap which persists as consumers' inability to translate the available information into sustainable consumption practice (Pape et al., 2011; Valor, 2008).

Methodology

The purpose of this paper is to evaluate how consumer attitudes and behaviours affect participation in sustainable garment disposal. In order to do this, it was necessary to establish an understanding of how consumers purchase and discard fashion and textile products, as well as current consumer attitudes and behaviours in relation to the design, promotion and retail of sustainably produced fashion. An explanatory sequential mixed methods approach was taken, which allowed the study to expose and understand the underlying mechanisms at work behind these structures. The UK was chosen as a representative European country for the first phase of the research, in which a quantitative data exposed behavioural patterns. Consumer focus groups in the second phase of the research enabled qualitative data to explain and provide insight on the drivers of these behaviours (Subedi, 2016).

Literature reviewed indicates that a major barrier to the widespread uptake of circular economy strategies is the 'values–action gap', which exists when consumers express ethically motivated intentions, but fail to follow this up with behaviour reflective of their concerns. Literature also indicated that understanding consumer involvement, knowledge and product literacy are key to promoting behaviour change. In order to investigate the complexities behind these consumer dimensions, it was necessary to consider the perspectives of individual citizens and their values, intentions and behaviours, as well as the sources of their knowledge and information.

Phase 1: Research design and sampling

In the first stage of this study, an online questionnaire was developed to gather data on consumer attitudes and behaviours. In 2015, a sample of 353 UK based consumers with an interest in fashion shopping was accessed through internet based social networks and the survey distributed through online snowball sampling. The survey was deemed suitable for gathering standardised, quantitative data, with predominantly closed questions (Denscombe, 2010; Bryman, 2012). In order to establish an overarching perspective of current consumer behaviours and practices regarding involvement with sustainable disposal habits, the source of consumer knowledge and the extent of sustainable product literacy, survey questions collected data on demographic categories, psychographic characteristics, behaviour motivators and sustainable fashion consumption attitudes and divestment behaviours.

Phase 2: Research design and sampling

In the second stage of this study, focus groups were used to gather data from European citizens, in order to identify the drivers of their attitudes and behaviours regarding use, disposal, and engagement with information to support sustainable practices. Between October 2017 and April 2018, focus groups were carried out as part of the EU Resyntex project's 'Citizen Labs' in four case study areas of Manchester (UK), Prato (Italy), Annecy (France) and Maribor (Slovenia), as detailed in Table 1.

In organising the focus groups, organisations such as schools, museums, universities and private companies based locally in the case

Table 1. Numbers of participants in case study location

Case study	Manchester	Maribor	Annecy	Prato
Number of participants	36	51	23	63
Average age of participants[1]	26	22	30	20
Gender distribution	34/2 (F/M)	39/12 (F/M)	14/9 (F/M)	22/41 (F/M)

[1] Calculated on the basis of information which was voluntarily disclosed, and therefore not exhaustive.

study region were partnered with as case study sample gatekeepers. This enabled the focus group participants to be purposively sampled as the target demographic. Partner organisations hosted a Resyntex Citizen Lab and mobilised their network of students, employees, visitors, partners, clients, etc. to participate in a focus group. This format allowed for a broad participation, enabling an engagement with ordinary consumers who may or may not have an affinity with the issue of sustainability and waste management. The Citizen Labs had a duration of around 1.5 hours and consisted of a online survey part which built on the UK survey, to extend the insights established, and of a series of interactive sessions whereby participants were encouraged to discuss and share their views and experiences, and wherein they actively sorted and discarded waste textiles and were enabled to explain the motivations and perceptions steering their choices.

The themes of inquiry centred around the key topics for investigation established from literature:

- Product Literacy: Fashion, shopping and perception of products (sustainable, recycled etc.).
- Consumer Involvement: Disposal, donation and use phase drivers.
- Consumer Knowledge: Preferred information sources – on disposal/donation etc. and on fashion/shopping.

Data analysis

Quantitative data were analysed using descriptive statistics to characterise the sampled participants with an indication of which responses were most frequent and how these responses were distributed amongst respondent groups. When appropriate statistical techniques such as correlation analysis, analysis of variance and cross-tabulation were carried out in SPSS software. This enabled examination of the strength of relationship between two variables, variance between groups and within groups, and interrogation of the data for patterns of association between variables. Qualitative data was thematically analysed through a system of coding.

The data were broken down into key units of analysis that represent what was said by participants. Each unit was coded and categorised according to significance to the research goals. Coded data units indicated classes of behaviours, from which conclusions can be drawn.

Findings and Discussion

Survey findings firstly presented insights into behavioural patterns which led to disposal habits. Focus group results then presented insights into the drivers of these behaviours, including preferred sources of information, exposing the level of consumer involvement with sustainable garment disposal practices.

Phase 1: Survey

Conditions for involvement

60% of respondents (n=213) claimed that ethical and environmental issues are 'Often' and 'Always' important to them, indicating an awareness of the importance of these issues, however only 21% (n=77) indicated that they had 'Often' and 'Always' purchased clothing because of the ethics of the brand making it. The gap of 39% of survey respondents who are yet to translate their concerns into more responsible behaviour represent the 'values-action gap' reported at 30% by Young et al. (2010).

When asked to rank fashion industry stakeholders in terms of responsibility for ethical and environmental choices, Table 2 indicates that survey respondents viewed themselves as customers to have the least responsibility, and designers, brands and shops the most, followed

Table 2. Mean scores for responsibility rankings by consumers

Rank	Mean score	Stakeholder group		
1	3.78	Fashion Designers, Retailers, Brands and Shops	MOST RESPONSIBLE	
2	3.05	Factories and Employers		
3	3.00	The Government		
4	2.63	The Media		
5	2.54	Customers	LEAST RESPONSIBLE	

by factories and employers, the government and the media. These findings indicate that consumers expect to have ethical and environmental considerations taken care of by those selling them their clothes, and that as consumers they feel little responsibility for making ethical choices themselves. These first stage survey respondents indicated that they viewed brands and retailers as having the greatest responsibility for making conscientious ethical and environmental choices, such as end-of life garment considerations within their own brand policies.

Consumer involvement: Garment use and divestment

The majority of survey respondents reported dealing with worn out clothes by recycling them at home into cleaning cloths (29%, n=101), taking them to a recycling bank (28%, n=97) or throwing them in the bin (26%, n=93), as shown in Figure 2.

Findings suggest low involvement with sustainable disposal behaviours, as informed disposal practices are perceived to be of minimal relevance to participants (O'Cass, 2000). While consumers may feel that they are participating in sustainable behaviours by recycling unwanted items into rags, these items will inevitably end up in the bin. As a significant proportion is also reported to be disposed off directly into the bin, these outcomes provide evidence that the majority of unwanted, worn-out clothing and textile items eventually enter into municipal solid

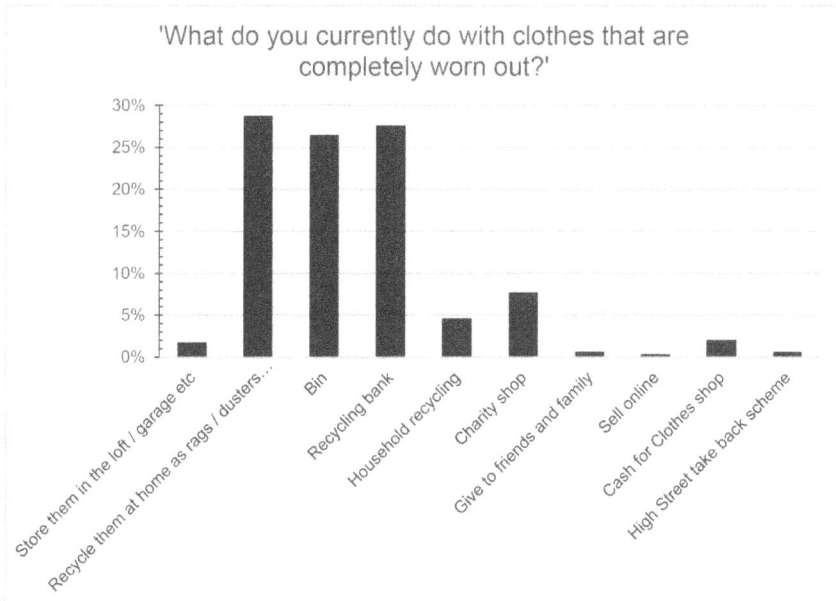

Figure 2. Ways of dealing with clothes considered to be 'worn out'

waste. Socks and underwear were overwhelmingly disposed off into the bin (70%, n=248), with some recycled as rags or taken to a recycling bank.

However, clothes that first stage respondents were bored of or did not fit anymore were mostly taken to a charity shop (~50%), but some were also stored in lofts and garages (~15%), given to friends and family (~13%) or sold online (~10%). This demonstrates that survey participants felt there to still be some embodied value left in these items and is an indicator of the opportunity for more high involvement practices.

Over ¼ (26%, n=93) of all first stage survey respondents placed worn out clothing in the bin. Rags and cleaning cloths will also eventually make their way into municipal waste streams. As survey respondents reported often binning items, this suggests that involvement is indeed inhibited by a lack of in-depth product knowledge and familiarity with end of use value.

The main reasons given for throwing clothes and textiles in the bin (Figure 3) were that respondents felt these items to be too worn out or dirty to be recycled (60%, n=210), or thought that they would not be worth anything to a charity shop (39%, n=136). This is an indication that consumers are lacking the information, knowledge and understanding of how these items could be valued and reused in a circular economy fashion and textiles system. This is confirmed by Gracey and Moon (2012) who show that the main reason given for throwing clothes in the bin is that consumers believe they could not be used again for any purpose (75%), followed items being too personal to get rid of another way (37%) and then a belief that the items no longer have any monetary value (26%).

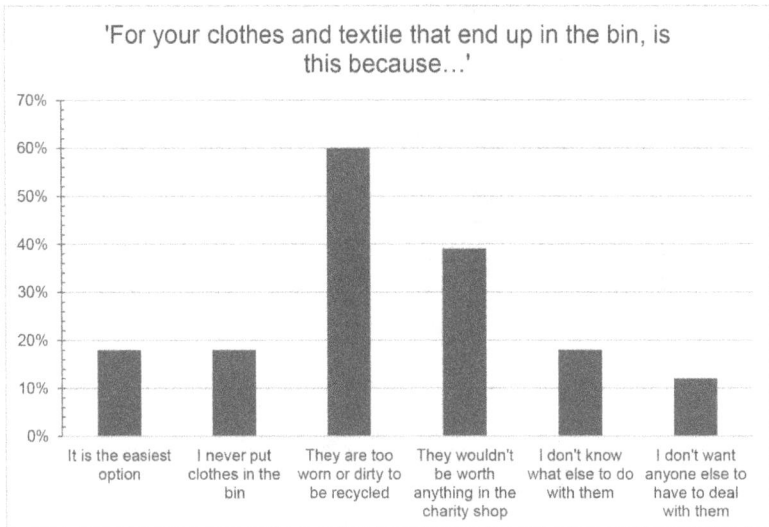

Figure 3. 'For your clothes and textiles that end up in the bin, is this because...'

For clothing that gets binned as it is considered too damaged or worn, it is predicted that more would be separated out for reuse and recycling if individuals were made aware that these items still represented a commercial reuse value as secondary raw materials (Gracey and Moon, 2012). An opportunity is presented by these findings to communicate clearer and more accessible information to consumers on how the clothes and textiles they had regarded as waste could be collected and re-valued. Statistical analysis revealed that those in the 35 to 44 age group were most likely to use a textile recycling bank, with the youngest age group the least likely to do so. Gracey and Moon (2012) reported similar findings that 37% of consumers used a textiles bank to get rid of their unwanted clothes, and Bartlett et al. (2013) also report that 36% of textiles are collected through this route. 62% (n=218) of consumers surveyed would often or always donate their worn out or broken clothes to the charity shop. This was also shown to be the most popular option by Gracey and Moon (2012) and the largest collection route for textiles by Bartlett et al. (2013).

As shown in Figures 4 and 5, first stage respondents were surveyed on the convenience and time taken while donating clothes responsibly, either to a charity shop or by taking them to a textile bank.

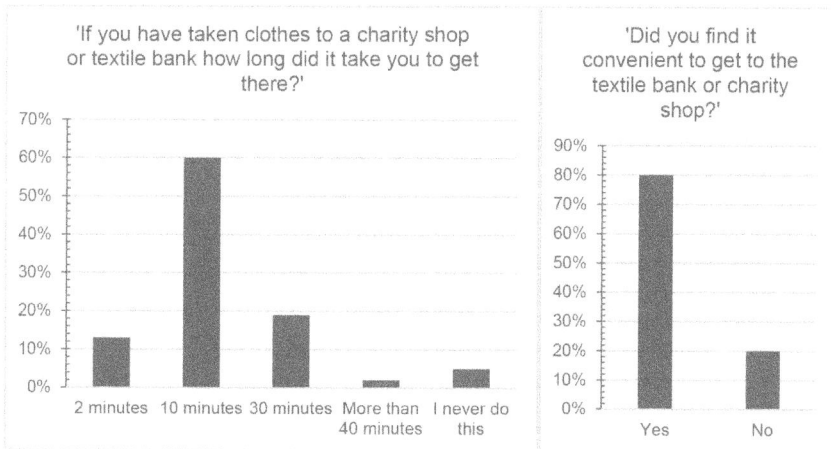

Figures 4 and 5. 'Time taken for donation' and 'Convenience'

It can be seen in Table 3 that the percentage of those finding donating convenient is higher as less time was taken to donate. The most convenient time is 2 minutes, with 97.9% (n=46) of donators finding this agreeable, however the most common time taken for clothing donations is 10 minutes, still with 91% (n=193) of donators finding this convenient. For those not finding donating convenient and spending over 30 minutes (47.1%, n=32) or 40 minutes (57.1%, n=4) to reach a donating destination, creating closer and easier to reach points of textile collection may increase the amount of textiles which could potentially be collected.

This established proximity and convenience is a key consideration in optimising used textile collection as part of circular economy fashion system. Sidique et al. (2010) established that as the round trip distance from home to recycling site increases per mile, the number of visits to the site decreases by 1%, confirming extant research by Saphores et al. (2006) that closer proximity encourages recycling behaviour. Convenience is an over-riding factor in high involvement behaviours.

Table 3. Convenience and donation of clothes and textiles

Did you find it convenient to get to the textile bank or charity shop?		If you have taken clothes to a charity shop or textile bank how long did it take you to get there?				
		2 minutes	10 minutes	30 minutes	More than 40 minutes	I never do this
Yes	Number	46	193	36	3	4
	Row	16.3%	68.4%	12.8%	1.1%	1.4%
	Column	**97.9%**	**91.0%**	**52.9%**	**42.9%**	**21.1%**
No	Number	1	19	32	4	15
	Row	1.4%	26.8%	45.1%	5.6%	21.1%
	Column	2.1%	9.0%	47.1%	57.1%	78.9%
X^2	103.5	df	4	Sig.	0.000***	

*** Refers to statistical significance

Consumer knowledge: Influences and information

Literature suggests that external barriers to scaling up circular economy fashion practices include a lack of market knowledge relating to consumers and the most effective strategies to connect with them. Survey insights show the youngest demographic group to be the most characteristic fashion leaders, but with the least regard for conscientious consumption and divestment. Figure 6 illustrates that around 60% of all respondents learnt what to do with old clothes and textiles while growing up at home, 37% while talking with friends and family and 20% from flyers through the door. This confirms the influence of upbringing, family life and peers on the consumer behaviour and decisions of children and young people (Joung and Park-Poaps, 2013). Findings show younger respondents expressed the strongest preference for receiving information through social interaction, either online or in person, using social media and blogs, or talking or shopping with friends and family. In order to connect with the younger fashion leaders and fashion followers, using socially engaging communication through social media, online content and shared peer and family experiences would yield the most effective results in promoting responsible divestment options for circular economy fashion.

'Where do you find information on what to do with your old clothes?'

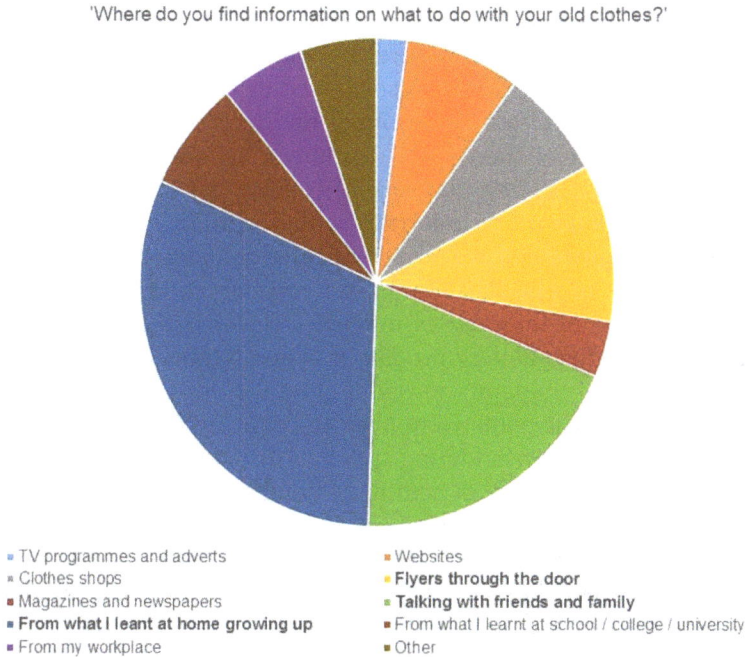

- ■ TV programmes and adverts
- ■ Clothes shops
- ■ Magazines and newspapers
- ■ **From what I leant at home growing up**
- ■ From my workplace
- ■ Websites
- ■ **Flyers through the door**
- ■ **Talking with friends and family**
- ■ From what I learnt at school / college / university
- ■ Other

Figure 6. 'Where do you find information on what to do with your old clothes?'

While most participants indicated high levels of objective product knowledge (Park et al., 1994) in relation to trend and style preferences, the concept of subjective product knowledge in relation to product disposal was more prevalent in their responses. Subjective knowledge was informed predominantly by the habits learned within the household, the domestic setting and via word of mouth amongst peers and family members. Little evidence has been found regarding the development of objective knowledge of sustainable disposal behaviours via mediated sources of information.

Phase 2: Focus groups

Survey findings present significant evidence that an understanding of the antecedents to involvement and product literacy are critical in the development of effective consumer engagement with behaviours and practices to support a circular economy for fashion and textiles. Consequently, focus group discuss insights into the drivers of these behaviours, including preferred sources of information, exposing the level of consumer involvement with sustainable garment disposal practices.

Consumer involvement: Disposal behaviours

Consistent with survey findings, when discarding garments, the most frequently mentioned routes were the waste bin, charity and second-hand shops, and textile banks. Also aligned to survey findings, there are two main factors driving determining the decision-making in relation to the destination of these textile products: quality and convenience. Indeed, during the focus group sorting exercise, participants consistently explained their choices by referring to the qualitative condition of the product being sorted rather than the material contained within it. Hence, the assessment of potential reuse and recyclability is primarily done through the lens of the product in itself – it was found that consumers do not consider the value or process of material (fibre) recycling.

Through more in-depth questioning on the use of textile banks and charity shops, participants were keen to confirm that convenience was key to the enabling or hindering factor of disposal. 80% of participants had, in the past, donated an item of clothing to a charity shop or textile bank, of which a very large majority indicated it was convenient to do so. The primary motivations were ethical (giving to those in need) and environmental (enabling a reuse of the product). The lack of local textile banks, as well as the lack of information and transparency on what subsequently happens to their clothes, were cited as the main reasons for low involvement in sustainable disposal behaviours, i.e. not bringing their clothes to these dedicated collection points.

In response to the question: "What do you do with those items that are leaving your wardrobe?" other disposal routes were exposed. The younger respondents (aged 15-25) frequently sell items of clothing online or exchange them through friends or social media platforms. Knowledge and awareness of other disposal options, such as retailer take-back schemes and clothes swaps, recovering textiles for use as cleaning rags, and putting textiles in the mending box, appeared considerably lower. The Manchester group demonstrated interest in repurposing, with 26% responding positively. Examples are: "reuse it as household rags", "I have had clothes remodelled by a dress maker", "repurposing, for example cutting the legs off and wearing as shorts", "I experiment with clothes to see what I could make of it".

Findings corroborate those of the survey and those of Joung and Park-Poaps (2013), prompted by evidence from Goodwill Industries (Solid

Waste District of LaPorte County, 2008); convenience was again found to be an overriding factor affecting clothing disposal behaviours.

Information sources

Again, in alignment with survey results, the primary sources of information were reported as each individual's network of family and friends (word of mouth), in addition internet communications and implicitly through the presence and observation of textile containers and second-hand shops in their neighbourhoods. There was a consistent request to be informed more through social media and text messaging. The participants were also keen to receive advice on public campaigns to raise awareness on the issue of textile recycling. Possible means of achieving this were expressed as via billboard advertisements, the involvement of celebrities such as singers and sportsmen, as well as through communications by municipal councils and schools at the local level.

Product literacy: Shopping behaviours

As proposed by Kopp (2012), product literacy is the degree to which consumers have the capacity to locate, obtain, evaluate, apply and communicate basic information needed to make appropriate product related decisions. A very consistent result across all focus group outcomes was that the purchasing of clothes significantly outweighed any consideration of disposing of them. It was found that consumers tend to hold on to their items, even when no longer deemed of use confirming that the purchasing cycle moves at a much higher speed than the disposal cycle. Participants confirmed that they tended to locate, obtain and evaluate factors of price and quality at the point of purchase. The application and communication of price and quality related messages far outweighed their consideration of, for example, whether fashion products were produced locally or what fibre types were used to manufacture the product. Interest in organic cotton, fair trade manufacture, sustainability credentials, second-hand clothing and upcycled garments was low. According to Kopp (2012) this limitation in involvement and product literacy would inhibit participants motivation to make sustainably significant purchase choices.

The word "fashion" was routinely used when referring to acquiring garments (whether new or second-hand). It was important for participants to be "in fashion". However, the term "fast fashion" was not part of the vocabulary of the participants. This finding is significant, as it indicates that the participants in all four countries were not typically fast fashion consumers. Gabrielli et al. (2013) found that fast fashion shoppers were influenced by four factors: a low price, a low but acceptable level of quality, a broad assortment of designs and a frequent renewal of collections. They found that low quality expectations contributed to the attitude that fast fashion products are disposable.

In most cases, "recycled clothing" was used to describe garments with recycled fibre content, rather than referring to reused, reworked or repurposed garments. The business model of garment rental had almost zero response. Limited interest in alternative business models was also reported by Gwozdz and Nielsen (2017) and Gwozdz et al. (2017). In the past, it has been well-documented that, even when there are positive attitudes to ethical and environmental values, "these interests seldom translate into action" (Manchiraju and Sadachar, 2014).

It became apparent that the language of the Circular Economy was new to all the participants, and responses to focus group discussions were mainly expressed as questions. Thomas's (2008) warnings about the need for words to have clarity of meaning are relevant. Markkula and Moisander (2012, p. 118) "argued that part of the 'knowledge-to-action' gap [. . .] can be attributed to a discursive confusion that arises from a simultaneous existence of multiple, continuously changing and partly clashing discourses on sustainable consumption." Evidence confirmed that product literacy and the ability to make informed disposal decisions was being hampered by limited access to sufficient information about the Circular Economy.

Conclusions and recommendations

The focus group findings demonstrated that only few consumers have any appreciation of circular economy concepts, possess limited product knowledge beyond price and quality which leads to low involvement with sustainable disposal behaviours. In terms of antecedents to involvement, many of the words that are routinely used by advocates of the circular economy have limited relevance and meaning in the minds of consumers: the term "recycled" is one of these.

Respondents expressed major concerns about ethical and environmental issues but displayed a clear 'values-action gap' in translating these concerns into actual behaviour. Instead of committing themselves to acquire and dispose off products responsibly, they placed the greater responsibility on brands and designers. Consumers, in fact, find it easier to understand "circularity" when it refers to "fibre to fibre" textile recycling processes, as reported by Vehmas et al. (2018). In creating effective communications which fully encompass the industrial symbiosis involved in circular fibre to fibre reprocessing, developers must be mindful of consumers' current levels of product literacy.

To improve product literacy and facilitate greater consumer involvement with sustainable disposal practices, findings indicate that information about ethical and sustainability issues should be integrated into mainstream fashion information as standard practice. This "producer responsibility" stance should provide a pathway to enable consumers to develop informed and objective product knowledge which they can use to inform responsible purchase and disposal behaviours. Findings support the recommendations of Zane et al. (2015) that companies who wished to make the ethical credentials of their products a selling point, need to have this information easily accessible and freely available, so that all consumers are able to develop full product literacy. Consumers may have previously associated negative feelings with ethical consumption choices. However, effective communication of sustainable practices enables all stakeholders to participate in shared accountability. This is vital to ensure consumer commitment to a shared responsibility of ethical and environmental choices (Zane et al., 2015).

The research reported here reinforces the view that whilst consumers are sympathetic towards ethical and environmental perspectives, there needs to be a strong lead from producers and brands to educate and inform consumer to enhance familiarity with the circular concept and encourage involvement with sustainable product disposal. The language of sustainability needs to be used carefully, to avoid confusion in the minds of consumers. As proposed by Kopp (2012), product literacy will only be developed when consumer understand the net benefit of obtaining information about the disposal of their garments. Without the means to evaluate this information, consumers are unable to determine the costs and benefits of new disposal habits. Alternative approaches to disposal need to be piloted, knowing that much of Europe is already committed to the separate collection of textile wastes. Governments have a role here: via Extended Producer Responsibility regulation of all textile products offered to consumers. Although projects, such as Resyntex, have demonstrated that technical solutions are possible, the work needs to be continued so that circular economy practices become viable and standard practice.

References

Ajzen, I. (1985). From intentions to actions: A theory of planned behavior. *In:* J. Kuhl and J. Beckmann (Eds.), *Action Control: From Cognition to Behavior*. pp. 11-39. Berlin, Heidelber, New York: Springer-Verlag.

Ajzen, I. (1991). The theory of planned behavior. *Organizational Behavior and Human Decision Processes,* 50(2): 179-211.

Bartlett, C., McGill, I. and Willis, P. (2013). *Textiles Flow and Market Development Opportunities in the UK*. Waste & Resources Action Programme: Banbury, UK.

Beard, N. (2008). The branding of ethical fashion and the consumer: A luxury niche or mass-market reality? *Fashion Theory*, 12(4): 447-468.

Bray, J.P., Johns, N. and Kilburn, D. (2011). An exploratory study into the factors impeding ethical consumption. *Journal of Business Ethics*, 98(4): 597-608.

Bryman, A. (2012). *Social Research Methods*, 4th ed. Oxford: Oxford University Press.

Chartrand, T.L. (2005). The role of conscious awareness in consumer behavior. *Journal of Consumer Psychology*, 15(3): 203-210.

Connell, K.Y.H. (2010). Internal and external barriers to eco-conscious apparel acquisition. *International Journal of Consumer Studies*, 34(3): 279-286.

Dahlbo, H., Aalto, K., Eskelinen, H. and Salmenperä, H. (2017). Increasing textile circulation—Consequences and requirements. *Sustainable Production and Consumption*, 9: 44-57.

Denscombe, M. (2010). *The Good Research Guide*, 4th ed. Maidenhead: Open University Press.

Domina, T. and Koch, K. (1999). Consumer reuse and recycling of post-consumer textile waste. *Journal of Fashion Marketing and Management: An International Journal*, 3(4): 346-359.

Ellen MacArthur Foundation (2017). A new textiles economy: Redesigning fashion's future, https://ellenmacarthurfoundation.org/a-new-textiles-economy

Ellen, P.S. (1994). Do we know what we need to know? Objective and subjective knowledge effects on pro-ecological behaviors. *Journal of Business Research*, 30(1): 43-52.

Ellis, J.L., McCracken, V.A. and Skuza, N. (2012). Insights into willingness to pay for organic cotton apparel. *Journal of Fashion Marketing and Management: An International Journal*, 16(3): 290-305.

Field, A. 2013. *Discovering Statistics using IBM SPSS Statistics*, 4th ed. London: Sage Publications Ltd.

Fisher, T., Cooper, T., Woodward, S., Hiller, A. and Goworek, H. (2008). Public Understanding of Sustainable Clothing: A Report to the Department for Environment, Food and Rural Affairs, Defra. London.

Foxall, G.R., Goldsmith, R.E. and Brown, S. (1998). *Consumer Psychology for Marketing*, Vol. 1. Cengage Learning EMEA.

Gabrielli, V., Baghi, I. and Codeluppi, V. (2013). Consumption practices of fast fashion products: A consumer-based approach. *Journal of Fashion Marketing and Management: An International Journal*, 17(2): 206-224.

Goworek, H., Fisher, T., Cooper, T., Woodward, S. and Hiller, A. (2012). The sustainable clothing market: An evaluation of potential strategies for UK retailers. *International Journal of Retail & Distribution Management*, 40(12): 935-955.

Gracey, F. and Moon, D. (2012). *Valuing Our Clothes: The Evidence Base*. Waste & Resources Action Programme (WRAP). Available online: http://www. wrap. org. uk/sites/files/wrap/10.7, 12.

Gwozdz, W., Steensen Nielsen, K. and Müller, T. (2017). An environmental perspective on clothing consumption: Consumer segments and their behavioral patterns. *Sustainability*, 9(5): 762.

Gwozdz, W., Nielsen, K.S., Gupta, S. and Gentry, J. (2017). *The Relationship between Fashion and Style Orientation and Well-being*. Mistra Future Fashion.

Harris, F., Roby, H. and Dibb, S. (2016). Sustainable clothing: Challenges, barriers and interventions for encouraging more sustainable consumer behaviour. *International Journal of Consumer Studies*, 40(3): 309-318.

Hobson, K. (2001). Sustainable lifestyles: Rethinking barriers and behaviour change. *Exploring Sustainable Consumption: Environmental Policy and the Social Sciences*, 1: 191-209.

Hobson, K. (2003). Thinking habits into action: The role of knowledge and process in questioning household consumption practices. *Local Environment*, 8(1): 95-112.

Houston, M.J. and Rothschild, M.L. (1977). *A Paradigm for Research on Consumer Involvement*. Graduate School of Business, University of Wisconsin-Madison.

Hustvedt, G. and Bernard, J.C. (2008). Consumer willingness to pay for sustainable apparel: The influence of labelling for fibre origin and production methods. *International Journal of Consumer Studies*, 32(5): 491-498.

Jackson, T. and Michaelis, L. (2003). *Policies for Sustainable Consumption*. Sustainable Development Commission, London.

Jackson, T. (2004). Consuming paradise? Unsustainable consumption in cultural and social-psychological context. *In:* Hubacek, Klaus, Atsushi Inaba and Sigrid Stagl (Eds.), *Driving Forces of and Barriers to Sustainable Consumption*, Proceedings of an International Conference, University of Leeds, 5th-6th March 2004.

Johnson, E.J. and Russo, J.E. (1984). Product familiarity and learning new information. *Journal of Consumer Research*, 11(1): 542-550.

Josiam, B.M., Kinley, T.R. and Kim, Y.-K. (2005). Involvement and the tourist shopper: Using the involvement construct to segment the American tourist shopper at the mall. *Journal of Vacation Marketing*, 11: 135-154.

Joung, H.-M. and Park-Poaps, H. (2013). Factors motivating and influencing clothing disposal behaviours. *International Journal of Consumer Studies*, 37(1): 105-111.

Kang, J. and Park-Poaps, H. (2011). Motivational antecedents of social shopping for fashion and its contribution to shopping satisfaction. *Clothing and Textiles Research Journal*, 29(4): 331-347.

Kang, J., Liu, C. and Kim, S.H. (2013). Environmentally sustainable textile and apparel consumption: The role of consumer knowledge, perceived consumer effectiveness and perceived personal relevance. *International Journal of Consumer Studies*, 37(4): 442-452.

Kim, H.-S. and Hong, H. (2011). Fashion leadership and hedonic shopping motivations of female consumers. *Clothing and Textiles Research Journal*. 29(4): 314-330.

Kinder, L.E. (2014). Fair trade apparel: Purchase intent among young female adults. *In:* Hutton Honors College Indiana University-Bloomington 2014 Undergraduate Research Symposium & Fair.

Kopp, S.W. (2012). Defining and conceptualizing product literacy. *Journal of Consumer Affairs*, 46(2): 190-203.

Koszewska, M. (2013). A typology of Polish consumers and their behaviours in the market for sustainable textiles and clothing. *International Journal of Consumer Studies*, 37(5): 507-521.

Laaksonen, P. (1977). *Consumer Involvement: Concepts and Research*. London: Routledge.

Laitala, K. (2014). Consumers' clothing disposal behaviour – A synthesis of research results. *International Journal of Consumer Studies*, 38(5): 444-457.

Manchiraju, S. and Sadachar, A. (2014). Personal values and ethical fashion consumption. *Journal of Fashion Marketing and Management*, 18(3): 357-374.

Markkula, A. and Moisander, J. (2012). Discursive confusion over sustainable consumption: A discursive perspective on the perplexity of marketplace knowledge. *Journal of Consumer Policy*, 35: 105-125.

McEachern, M.G., Warnaby, G., Carrigan, M. and Szmigin, I. (2010). Thinking locally, acting locally? Conscious consumers and farmers' markets. *Journal of Marketing Management*, 26(5-6): 395-412.

McNeill, L. and Moore, R. (2015). Sustainable fashion consumption and the fast fashion conundrum: Fashionable consumers and attitudes to sustainability in clothing choice. *International Journal of Consumer Studies*, 39(3): 212-222.

Michaelidou, N. and Dibb, S. (2006). Product involvement: An application in clothing. *Journal of Consumer Behaviour: An International Research Review*, 5(5): 442-453.

Mintel (2008). *Ethical and Green Retailing*. Mintel International Group Limited, London.

Mintel (2016a). Clothing Retailing - UK - October 2016: Attitudes Towards Buying.

Mintel (2016b). Womenswear - UK - May 2016: Most Important Factors When Buying Clothes.

Morgan, L.R. and Birtwistle, G. (2009). An investigation of young fashion consumers' disposal habits. *International Journal of Consumer Studies*, 33(2): 190-198.

Naderi, I. (2013). Beyond the fad: A critical review of consumer fashion involvement. *International Journal of Consumer Studies*, 37(1): 84-104.

O'Cass, A. (2000). An assessment of consumers product, purchase decision, advertising and consumption involvement in fashion clothing. *Journal of Economic Psychology*, 21(5): 545-576.

Pape, J., Rau, H., Fahy, F. and Davies, A. (2011). Developing policies and instruments for sustainable household consumptions: Irish experiences and futures. *Journal of Consumer Policy*, 34: 25-42.

Pappalardo, J.K. (2012). Product literacy and the economics of consumer protection policy. *Journal of Consumer Affairs*, 46(2): 319-332.

Park, C.W., Mothersbaugh, D.L. and Feick, L. (1994). Consumer knowledge assessment. *Journal of Consumer Research*, 21(1): 71-82.

Pentecost, R. and Andrews, L. (2010). Fashion retailing and the bottom line: The effects of generational cohorts, gender, fashion fanship, attitudes and impulse buying on fashion expenditure. *Journal of Retailing and Consumer Services*, 17(1): 43-52.

Phau, I. and Lo, C.-C. (2004). Profiling fashion innovators: A study of self-concept, impulse buying and Internet purchase intent. *Journal of Fashion Marketing and Management*, 8(4): 399-411.

Reimers, V., Magnuson, B. and Chao, F. (2016). The academic conceptualisation of ethical clothing: Could it account for the attitude behaviour gap? *Journal of Fashion Marketing and Management: An International Journal*, 20(4): 383-399.

Ritch, E.L. (2015). Consumers interpreting sustainability: Moving beyond food to fashion. *International Journal of Retail & Distribution Management*, 43(12): 1162-1181.

Saphores, J.D.M., Nixon, H., Ogunseitan, O.A. and Shapiro, A.A. (2006). Household willingness to recycle electronic waste: An application to California. *Environment and Behavior*, 38(2): 183-208.

Scheufele, D.A. (1999). Framing as a theory of media effects. *Journal of Communication*, 49(1): 103-122.

Sender, T. 2011. *Women's Fashions Lifestyles*. Mintel International Group Limited, London.

Sidique, S.F., Lupi, F. and Joshi, S.V. (2010). The effects of behavior and attitudes on drop-off recycling activities. *Resources, Conservation and Recycling*, 54(3): 163-170.

Sisco, C. and Morris, J. (2012). *The NICE Consumer: Toward a Frame-work for Sustainable Fashion Consumption in the EU*. The Danish Fashion Institute.

Subedi, D. (2016). Explanatory sequential mixed method design as the third research community of knowledge claim. *American Journal of Educational Research*, 4(7): 570-577.

Sudbury, L. and Böltner, S. (2011). Fashion marketing and the ethical movement versus individualist consumption: Analysing the attitude behaviour gap. *European Advances in Consumer Research*, (9): 163-168.

Thomas, S. (2008). From "Green Blur" to ecofashion: Fashioning an Eco-lexicon. *Fashion Theory*, 12(4): 525540.

Tomolillo, D. and Shaw, D. (2003). Undressing the ethical issues in clothing choice. *International Journal of New Product Development and Innovation Management*, 15(2): 99-107.

Valor, C. (2008). Can consumers buy responsibly? Analysis and solutions for market failures. *Journal of Consumer Policy*, 31(3): 315-326.

Vehmas, K., Raudaskoski, A., Heikkilä, P., Harlin, A. and Mensonen, A. (2018). Consumer attitudes and communication in circular fashion. *Journal of Fashion Marketing and Management: An International Journal*, 22(3): 286-300.

Workman, J.E. and Cho, S. (2012). Gender, fashion consumer groups, and shopping orientation. *Family and Consumer Sciences Research Journal*, 40: 267-283.

Young, W., Hwang, K., McDonald, S. and Oates, C.J. (2010). Sustainable consumption: Green consumer behaviour when purchasing products. *Sustainable Development*, 18(1): 20-31.

Zaichkowsky, J.L. (1985). Measuring the involvement construct. *Journal of Consumer Research*, 341-352.

Zane, D.M., Irwin, J.R. and Reczek, R.W. (2015). Do less ethical consumers denigrate more ethical consumers? The effect of willful ignorance on judgments of others. *Journal of Consumer Psychology*, 26(3): 337-349.

The Impact of Modes of Acquisition on Clothing Lifetimes

Kirsi Laitala*, Ingun Grimstad Klepp and Lisbeth Løvbak Berg

Consumption Research Norway (SIFO), Oslo Metropolitan University
e-mail: *kirsil@oslomet.no

Introduction

Product lifetime is a growing research field within clothing and environmental studies. In this chapter we will discuss the limitations of this perspective: If an increased lifetime is going to contribute to reducing the environmental impact, it must affect how much is being bought. This is because reducing environmental impact is dependent on a reduction in the volume of clothing produced, consequently reducing the acquisition of new clothes, as well as reducing environmental impacts in re-use systems. The literature lacks insight into the relationship between lifetimes and the number of clothes that are being produced and bought. In this chapter, we will examine this connection by asking how acquisition impacts the lifetimes of clothing both in terms of the mode of acquisition as well as the volume.

Clothing lifetime can be measured by how many times a garment is used, how long it is used, and how many users it has (Klepp et al., 2020). Even though the use phase is the most important part of the clothing lifespan, the discussion around prolonging product lifetimes has largely been focused on the production phase and on what designers and producers can do. However, there is no simple connection between garments that are designed to be durable and those that are used for a long time (Fletcher, 2012). Furthermore, the impact of the acquisition phase is understudied, as most consumer studies on clothing lifetimes focus on the use and disposal phases (Bernardes et al., 2020; Domina and Koch, 1997; Klepp and Laitala, 2016a; Klepp et al., 2015; Lai and Chang,

2020; Laitala, 2014; Laitala and Klepp, 2015; Rathinamoorthy, 2020; Shim, 1995). The acquisition phase has been a specific topic in anti-consumption studies, but these do not again focus on the length of product lifetimes as such (Vesterinen and Syrjälä, 2022).

Disposal of a product has often been seen as the reason for the acquisition of a new one as a replacement (Cooper, 2010; van Nes and Cramer, 2006). This is at the core of the idea that a longer lifetime for one product will prevent another from being acquired – and thus reduce the impact of clothing consumption on the environment. Clothing is acquired for many reasons, like treating yourself to something new, something you don't already own, on impulse when finding something aesthetically pleasing or a bargain. There is a connection between acquisition and disposal, but it is complex. Maldini et al. (2019) describe it as a "push and pull system". When new items are coming in, they push the older, less desirable garments out, or temporarily to the back of the wardrobe. When something is worn out, it can on the other hand 'pull' something new into the wardrobe. This illustrates the argument that acquisition and disposal are two different processes for clothing, and both can potentially affect each other. Both flows also affect the amount of clothing in the wardrobe which in turn affects lifetimes. To better understand wardrobe metabolism, better knowledge of how acquisition influences the lifetime of clothing – and the total environmental impact – is needed.

This chapter is based on a scoping review of previous research as well as a re-analysis of international wardrobe audits conducted in China, Germany, Japan, the UK and the USA (Nielsen, 2019). The online questionnaire was answered by 1111 working-age adult respondents with 213–230 respondents from each country. It included questions about their wardrobe content and usage practices related to 53461 specific clothing items. To keep the length manageable, not all questions were stated to all garments, thus the sample size varies some in the analysis. Questions that were re-analysed for this chapter concern clothing acquisitions (modes and amounts), length of clothing lifespans and wardrobe sizes. More details of the method are given in previous publications (Klepp et al., 2020; Laitala and Klepp, 2020b).

Acquisition

Clothing acquisition is here understood as all forms of actions that can give a consumer access to use a garment. The access can be permanent, based on ownership such as when a garment is purchased or received as a gift, or temporary when it is rented or borrowed (Corrigan, 1989). These acquisition modes can happen outside or inside the pecuniary market, such as when something is swapped instead of bought. It can be more or less planned and the clothes can be self-made, made-to-order or ready-made on the market. Some garments are also acquired in less orderly manners, such as when someone forgets an item and the finder takes it in use, or through shop-lifting (Strathern, 2011). The extent of these various forms of acquisition has changed throughout history and varies between cultures and consumer groups. An example of such change is that trade with used clothing predates trade with ready-made items (Klepp and Tobiasson, 2021; Lambert, 2004).

The international wardrobe audit (Nielsen, 2019) shows that the most common way in contemporary affluent societies is that the respondents themselves purchase new ready-to-wear garments (74%) followed by receiving gifts (10%) (Figure 1). It is also quite common to acquire used clothing, either through private networks such as hand-me-downs from family or friends (4%) or by purchasing second-hand clothing either at physical stores, markets or online (5%). Some clothes were also custom-made for the users (1.8%), either by professionals such as tailors, or privately, either by the users themselves or by someone they know.

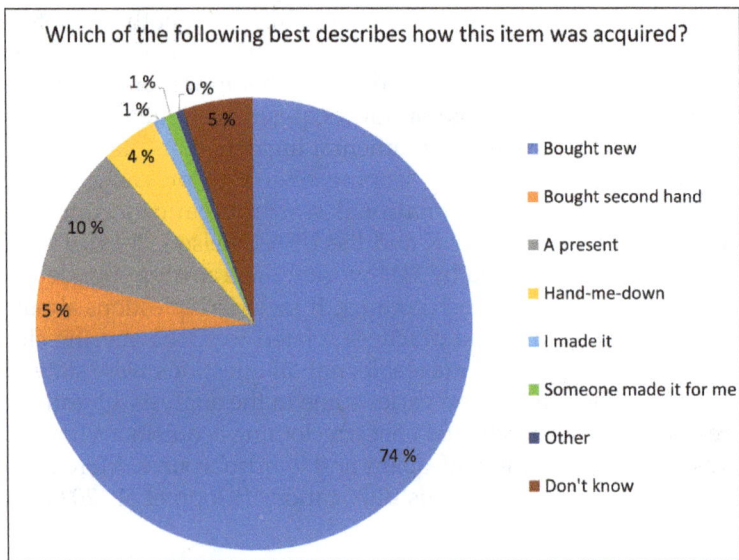

Figure 1. Modes of acquisition per garment (N=40356 garments)

How we acquire clothes also covers many other aspects than simply the mode of acquisition. We can acquire large or small quantities, expensive or cheap, and we can have different routines regarding frequency, how the acquisition is organised in time and how it is planned. We may also have different habits for different garment types, e.g., buying socks when the drawer is empty but dresses with the view of a special occasion even though our wardrobe contains several already. The acquisition can also have a ritual aspect to it, such as new clothing for starting school or related to pregnancy.

In this chapter, we will focus on the most important modes of acquisition, buying new or used clothing either online or at physical stores, personalised clothing, gifts and sharing, as well as the important difference between buying little and a lot. There are several other aspects connected to acquisition that are likely to have a great impact on lifespans, such as the price of garments, the economic situation of the user, social expectations as well as personal preferences (Jørgensen and Jensen, 2012; Connor-Crabb and Rigby, 2019; Gabrielli et al., 2013), but that we will not be able to discuss in this chapter.

Used or new

Acquisition of used clothing does not only comprise purchasing second-hand items, which has been researched the most, but also handing down, exchanging and other forms of privately circulating items without the exchange of money (Laitala and Klepp, 2017; Tinson and Nuttall, 2007). The exchanges can be made either directly from consumer to consumer or through intermediaries and various platforms. Extensive research has been done on used clothing connected to motivation and barriers to acquisition (Laitala and Klepp, 2018; Lang and Zhang, 2019) and also the type of garments (Cassidy and Bennett, 2012; de Wagenaar et al., 2022). Wardrobe studies indicate that the average share of used clothing in wardrobes varies between 6 and 17%, but with great national and individual differences (de Wagenaar et al., 2022; Klepp et al., 2020; Laitala and Klepp, 2020a; Maldini et al., 2017).

In the discussion about clothing and the environment, it is often assumed that buying used clothes completely or partly substitutes the acquisition of new clothes. This is called the replacement or displacement rate, which is defined as "the degree to which the purchase of second-hand clothing and household textiles replaces the purchase of similar new items" (Nørup et al., 2019, p. 1026). The replacement rate is zero if the item comes in addition to other purchases, and 100% if the item replaces completely a new item and is used for as long as a new item would be used. The replacement rate is an essential factor for the environmental benefits of reuse, as it indicates the production that can be avoided (Vadenbo et al., 2017). There are few studies on this but estimations vary between 28% and 90% (Castellani et al., 2015; Depop, 2022; Farrant et al., 2010, Nørup et al., 2019; Stevenson and Gmitrowicz, 2012). In a study of Norwegian sportswear, we found that hand-me-downs were an addition to and not a replacement for buying new clothes, both among those who bought the most and those who bought the least (Klepp and Laitala, 2018). It did not impact the number of purchases of new items, because the respondents that were well-off bought new items anyway, and those with financial constraints would have chosen not to participate in the activities rather than buying new items. A similar tendency was also found in studies in Africa where those with the lowest disposable income would not have bought anything if the cheap used clothes had not been available (Nørup et al., 2019).

The lifespans of second-hand garments are a less studied topic than the acquisition. Our research has indicated that second-hand garments are worn on an average 30% fewer times than garments that were acquired as new (Laitala and Klepp, 2021). On average, new garments were worn 82 times and second-hand garments 57.5 times. However, this ratio varies between different garment types. The difference is less for skirts and dresses where second-hand garments are worn only 17% fewer times than those acquired as new, while the difference is larger for formal wear such as suits where the difference was 52% (Figure 2). Second-hand garments that were bought or received as hand-me-downs were used equally many times.

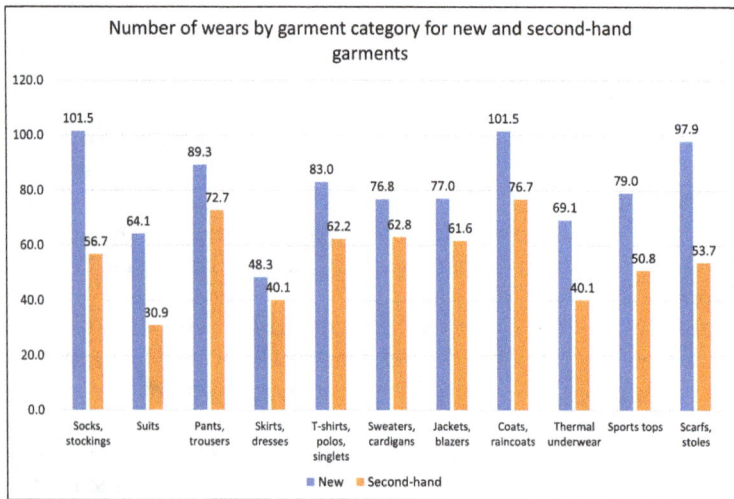

Figure 2. Number of wears by garment category for new and second-hand garments (new garments include self-purchased items, gifts and self-made and tailored items, while second-hand items include both purchased and hand-me-downs) (N=29820 garments)

Surprisingly, when measured in years, the difference is the opposite. Garments are kept 31% longer if they are second-hand. We do not know the reason for this. One possible explanation is that these items are kept as extra pieces and clothing for specific occasions more than everyday clothing and therefore will be used less – and last longer (Klepp et al., 2020). This question is also connected to the volume of clothing that is acquired, which we will discuss later.

Garments with more than one consequential user get several use phases and thus increased service life. The garment lifetimes measured in years more than doubled by having a new user, but the second user only increases the number of wears by 46% which is quite far from doubling the lifetime (Laitala and Klepp, 2021). This estimate has uncertainties as we only have information from one user and do not know how many users the pre-owned garments have had. In the Western peer-to-peer resale context, the users prefer to acquire pristine or barely worn goods at a price they could not ordinarily afford, which indicates short lifetimes with the first users of these items (Joyner Armstrong and Park, 2020). When measuring the replacement rate, it is important to study whether the used garment replaces a new purchase as well as the difference in use between a new and a used piece. Both of these factors are deciding when it comes to calculating the environmental advantages of reuse.

Online or in-store

The sales of new and second-hand clothing are increasingly moving online, and today, apparel is the most commonly purchased product group online (Eurostat, 2018, Statista, 2019). It has replaced the more traditional catalogue shopping in addition to a significant share of buying at physical stores (Cullinane and Cullinane, 2021) and now constitutes 40% of sales in Western Europe and 46% of sales in the USA (Berthene, 2021; Statista, 2022). Additionally, multichannel shopping is increasing where the consumers either search for information or inspiration online before visiting a physical store or go to a store to see and try on the items before ordering them online (Blázquez, 2014).

Private reuse is also increasing online, in 2021, 18% of European consumers sold (Eurostat, 2022a) and purchased goods (Eurostat, 2022b) online from other consumers. Despite this increase, Parker and Weber (2013) found that e-commerce such as eBay supplements and alters the markets rather than completely replaces the local thrift stores, vintage boutiques, and flea markets.

The increased availability of goods through online shopping is beneficial for consumers who either have special needs or otherwise limited access to clothing due to constraints based on time or location. However, when clothing is bought online, the possibilities for tactile cues such as feeling the fabric are missing (Citrin et al., 2003), as well as possibilities to try the garments on. A survey in Sweden showed that the main reason for the latest clothing return was that the item did not fit, given by 72% of respondents that had returned a clothing item during the past year (Cullinane et al., 2021). A large portion of clothing that is ordered online is bought with the intention of returning them afterwards, either through ordering a similar item in many sizes, colours or styles (IMRG,

2020), or after using them for short time, for example for photos in social media (Barclaycard, 2018).

We have not seen studies that compare clothing lifetimes purchased online or at physical stores, but several studies have shown that return rates of clothing purchased online are very high, varying from around 10% to over 60% for high fashion goods (Binkley, 2012; Cullinane et al., 2019; Cullinane and Cullinane, 2021; Roichman et al., 2023). This increases the environmental impact of transport significantly. Further, about 30% of returned items end up not being sold again (Reed, 2019), thus getting very short lifetimes related to no active wearing beyond being tried on. The average lifespan of clothing bought online is therefore reduced compared to those bought at stores. However, clothing bought online can be tried on at home with more time and space for movement than those tried on at stores. This enables a longer time for post-purchase consumer regret (Lee and Cotte, 2009), and therefore more careful consideration of the fit, appearance, and overall more thorough two-step selection process (first online and then at home) that may end up giving the items that are kept a longer use period. However, we lack evidence either way and based on the high return rates we would assume average lifetimes of online clothing are shorter than those bought in physical stores.

Personalised or ready-to-wear

The majority of garments today are mass-produced ready-to-wear items as opposite to customised/personalised clothing. Maldini et al. (2019, p. 1415) define personalised products in the apparel sector to be the

> *result of activities such as bespoke tailoring, made-to-measure, mass customisation, home or self-production and do-it-yourself. Despite the variety of practices involved, they all meet two central conditions that are not present in mass-produced ready-mades: individual user involvement in the design, and production on-demand.*

It is likely that this user involvement impacts the length of use times of these items, but there is more research on the motivation and the processes of craft than the use of personalised products (Holroyd, 2014).

The oldest garments in Norwegian wardrobes are most often woollen sweaters, which are also the most common homemade garments in Norway (Hebrok et al., 2016; Klepp and Laitala, 2016a, b). Another example of old garments is the Norwegian national costume which often includes tailored and homemade pieces (Klepp and Laitala, 2016b). Niinimäki and Koskinen (2011) studied consumers' oldest garments in Finland and found some similar tendencies, the perceived value of a garment can increase if it is personalised or has an emotional connection to a particular person or memory, in addition to functional and material qualities. Further, this value is a deciding factor in whether it is well taken care of and maintains its attractiveness over time (Niinimäki and Koskinen, 2011). A similar effect is also known from food research, where the relational value, the fact that someone we know has made the food, reduced food waste (Hebrok and Heidenstrøm, 2017).

Surprisingly, the wardrobe audit showed that the different categories of personalised items had very different lifetimes. Self-made clothing was worn the least number of times, while garments that were made by someone else (custom-made, tailored, made-to-measure etc.) were worn the most times (Figure 3). However, when we compare the age of self-made garments, we see that they are the second oldest group in our wardrobes, only beaten by custom-made items. We have previously seen that the more recently acquired garments are used more actively than the older garments in the wardrobe (Laitala and Klepp, 2021).

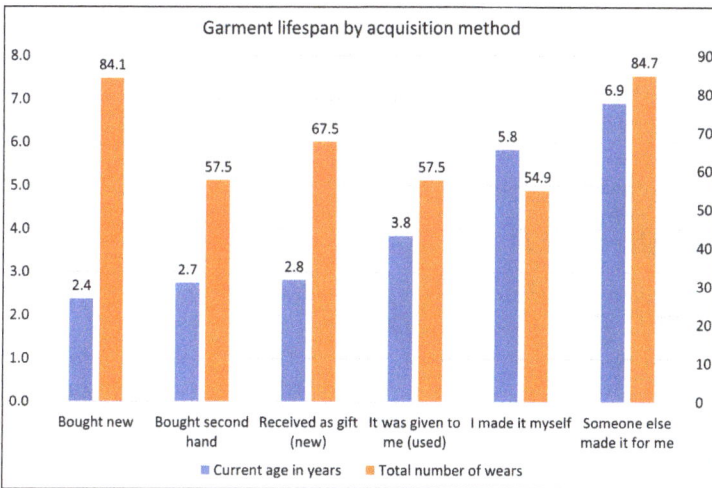

Figure 3. Age and the total number of wears of clothing acquired through different channels

We do not know why the lifetime measured in the number of wears is lowest for self-made clothes. Holroyd (2014) points out how homemade items differ from ready-to-wear apparel in many ways. While they lack the meanings associated with brands, they have multiple meanings either connected to positive associations of garments made with love, self-sufficiency and skills, or the opposite, negative associations connected to ill-fitting poorly made garments with the stigma of poverty (Holroyd, 2014). The skills of the maker seem therefore crucial for whether the garment will be used actively.

Another possible explanation is that those that make clothing have a lot of clothes and that their overall degree of clothing utilisation is low.

Maldini et al. (2019) found that personalised garments, both homemade and custom-made, were neither used more often nor kept for longer than ready-made garments. Furthermore, informants that owned more than ten personalised garments did not own fewer items in total, nor did they acquire fewer items than individuals owning only ready-mades. They found that informants that personalised their items reported an increase in wardrobe size after they started personalising, suggesting that this practice may play a role in encouraging materialism in this specific product category and therefore in growing clothing demand (Maldini et al., 2019).

The most discussed design strategy aimed at diminishing production volumes of clothing in sustainable fashion literature is user involvement in design and/or manufacture, but the strategy is based on conceptual explorations and not validated (Maldini and Balkenende, 2017). Our findings show that validation is needed. If we measure lifetime in years, the results are opposite than if we measure it in the number of wears. The assumption in the design strategies which is also underpinned by research on food, would be that self-made clothes are used more. The factors that seem particularly important here are the correlation between making or having it custom-made and the size of the wardrobe. This shows that there is a lot we do not know about home-production and other personalised clothing, including the impact of affection value and skills.

Gift or self-chosen

Wardrobe studies indicate that up to 9-10% of garments are gifts (Klepp et al., 2020; Laitala and Klepp, 2020a). Clothing given as gifts is a complex group. Clothes that we get at work and from organisations and events might be a form of advertising. Much of this is very impersonal and can also be of low quality and with elements that limit use, such as prints that refer to a specific company, event, place, or group. Gifts are also part of maintaining close relationships with other people (Mauss, 1970) and displaying care (Miller, 1998). Norwegians receive on average two garment gifts per year, but men and young people receive more gifts than women and people above the age of 30 (Laitala and Klepp, 2020a).

The recipients of gifts do not usually have control over what they will get unless they have been very specific about what they wish for. Therefore, the chance of garments not fitting the recipient's taste, body or activities is larger than in cases where the user chooses the products themselves.

Studies of Norwegian Christmas gifts show that clothes are the most common gift category, but 10% said they would not wear the clothes they received for Christmas (Bugge et al., 2019). We do not know how many decided to exchange these gifts. There is a high risk that these garments are not exchanged or used and thus are in danger of never being used and thus have no "lifetime" at all as we define this word. Wardrobe studies have shown that garments received as gifts are over-represented among the items that have never been used or were used only once or twice (Laitala and Boks, 2012). Further, the international wardrobe audit indicated that garments that were received as gifts were used less than those that the respondents had bought themselves as new, but more than second-hand garments (Figure 3). A study in Australia concerning the enduring/longevous garments in wardrobes indicated that only 4% of those still in use were gifts, while gifts constituted 7% of enduring/longevous garments that were kept but no longer used (Cramer, 2019), most likely due to their higher emotional value.

Gifts do not stand out as having a particularly high or low lifetime whether we measure it in years or the number of wears. A possible explanation is that this group of clothes contain very different types of garments. A closer analysis should therefore distinguish between company gifts, personal gifts and the more common purchasing of clothes for family members.

Shared or personal

Sharing can be a form of acquisition where the user gets access to the clothing without a change in ownership. There are various forms of providing this temporary access to use, such as renting, leasing, lending, borrowing and even "stealing" for example in form of covert borrowing from siblings (Corrigan, 1989; Klepp and Laitala, 2018; Tinson and Nuttall, 2007). Sharing and collaborative consumption has received increasing attention as potential strategies to enable a circular economy by activating idle assets. However, these strategies can cause unintended negative environmental impacts for example by fostering hyper-consumption by enabling constant wardrobe changes and therefore increasing carbon footprints (Henninger et al., 2021). Studies that have investigated the environmental impacts of rental systems in comparison with linear business models have shown that the clothing lifetimes in terms of use intensity and replacement rate of rental for purchased clothing are crucial for whether such models are preferable from an environmental point of view, in addition to transport issues (Johnson, 2020; Levänen et al., 2021).

Clothes are also lent out privately and mostly between those who live together (Corrigan, 1989; Klepp and Laitala, 2018). It is still rather unusual to rent or lease clothes for private use commercially (Gyde and McNeill, 2021), but more common in the professional markets where the businesses are responsible for providing the clothing either for their employees or clients, such as within the health sector or the military. In such cases, the users have less influence on the procurement, and they do not become the owners of the clothes.

Renting such items can increase lifetimes and reduce environmental impacts due to improved maintenance practices (industrial laundering and frequent repair), better fit (users can change items that do not fit), waste management and recycling practices, but also indirectly encourages renting companies to invest in durable materials (Kumar et al., 2022). Public procurement is an important part of consumption as it involves large quantities and because it is possible to create massive change through political decisions.

Consumer studies need to be complemented with studies of public procurement and other collective clothing consumption options to gain accurate knowledge of how sharing influences clothing lifetime. Consumer studies are less suitable, because few people rent clothes and because consumers will not have the information about lifetimes of rented items. The studies that have been done are few and concentrated on environmental impact more than lifetimes. The impact can increase or decrease depending on how the rental is organised, what kind of clothes are rented out and what they potentially replace (Henninger et al., 2021).

Many or few

There are big differences in how much clothing consumers acquire, own and dispose off. As could be expected, Figure 4 shows how wardrobe sizes increase with the increased acquisition. This indicates that the outflows do not follow the same speed as the inflows. In particular, those that acquired more than 50 garments in the past 12 months had large wardrobes. In Norway, studies indicate that people who buy a lot

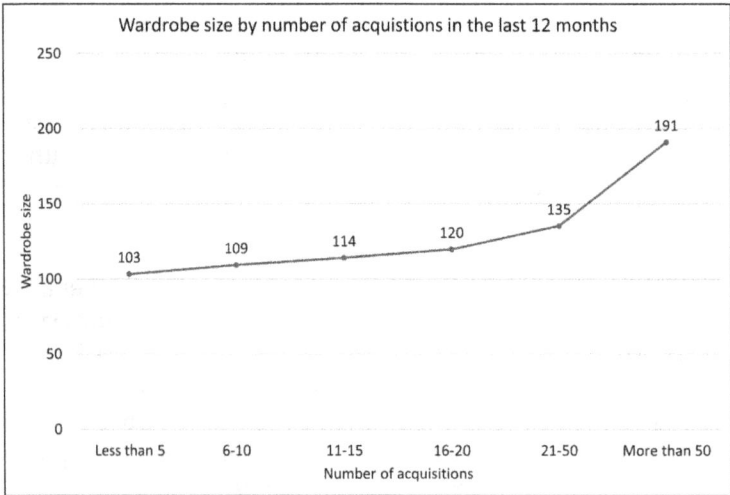

Figure 4. Wardrobe size by the number of acquisitions in the last 12 months (N=1042 respondents)

of second-hand clothing acquire a lot of clothes in general (Laitala and Klepp, 2020a).

Quantity is directly connected to lifetimes, as one can say that if consumers acquire less clothing, they will have to use what they have more. A lot of purchases will either lead to a larger wardrobe or a high turnover of clothing. With increased volumes, the clothes must be able to be used over several years if they are to be used as many times as in a small wardrobe. This is also confirmed by international wardrobe audit data that showed that consumers with large wardrobes use their clothes longer, but consumers with small wardrobes use their clothes more often before they are disposed off (Klepp et al., 2019). In a large wardrobe, clothing needs to age well, but in reality, many clothes have an expiry date whether in the form of an inflexible fit, the use of elastane that loses its elasticity, or more cultural and social aspects like fashion and design details that will affect use.

Figure 5 shows the connection between the number of acquisitions and lifetimes of clothing both in years and in the number of wears. Respondents that acquired less clothing in the past year used their existing clothes for longer and more frequently. The age of the garments is affected the most. The clothing belonging to those that bought the least had an average current age of 4.6 years and these clothes were estimated to be used in total 127.4 times (including future use). Those that bought the most, on the other hand, had clothing that was on average 1.9 years old and that they estimated to use 73.4 times. For the total lifetime, how they

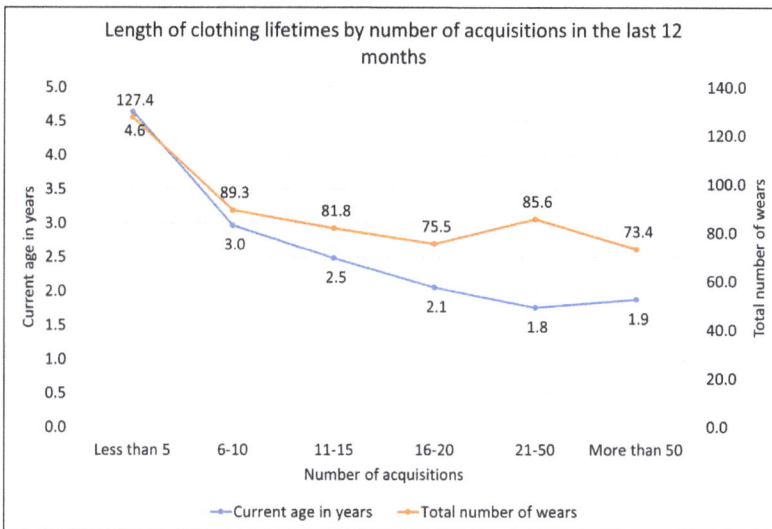

Figure 5. Length of clothing lifetimes by the number of acquisitions in the last 12 months (Years N=1036 respondents, wears N=966 respondents)

are disposed of will be important. This is especially valid for those that buy a lot and therefore discard clothes with remaining use-potential by a subsequent owner due to a lower level of wear and tear.

The link between acquiring few and many garments has an effect on the wardrobe size and through that, the lifetime both with the first and potential subsequent users. It is possible to utilise the clothes well – wear them a lot – both with a small and a large wardrobe, but the latter requires that they are used for a long time. Frequent purchases require a well-functioning second-hand market to increase clothing lifetime. Compared to the importance of purchase quantities for both environmental impacts and wardrobe sizes, there is very little research on the area.

Discussion: Utilisation rate

The examples from our analysis of global wardrobes and previous research show that the modes and quantities of acquisition affects clothing lifetimes. This relationship is complex and among other things depends on how the lifetime is measured. Variations in acquisition affected lifetime measured in years differently than lifetime measured in the number of wears. What influences the lifetime most directly is the amount of clothing acquired, and thus the amount of clothing in the wardrobe. A longer lifetime for one garment may just as well mean that another is used less rather than that it replaces a new purchase, as we discussed initially using the terms 'push' and 'pull'. This tells us that the total utilisation rate for the wardrobe is more important than the lifetime of each garment.

Ellen MacArthur Foundation (2017, p.24) argues that "increasing the average number of times clothes are worn is the most direct lever to capture the value and design out waste and pollution in the textiles system" and uses the term "clothing utilisation", meaning the number of times garments are used. This number has decreased for clothing, both in a centenary perspective and in the last 10 years, creating an increasingly inefficient industry, i.e., an industry that uses more and more resources to achieve the same result (a well-dressed population). This is a fact that the industry itself is aware of, for example, the Pulse report points out that "the number of times clothes are actually worn has dropped by a third compared to the early 2000s" (Lehmann et al., 2018, p. 59). Many numbers that are circulating on clothing utilisation are far below ours, such as that garments are worn only seven to ten times (Ellen MacArthur Foundation, 2017, p. 19) or that garments from specific brand are not designed to be durable but described as "to be worn 10 times" (McAfee et al., 2004, p. 4). However, many of these estimations have unclear origin or limited validity, such as studies limited to (young) female consumers in the UK (Daily Mail, 2015; Morgan and Birtwistle, 2009). Consequently, we lack reliable regional as well as global statistics on clothing utilisation.

The point of using the term clothing utilisation rather than lifetime is exactly that it points to our use of the clothes and not the clothes themselves. It is not the clothes that 'live' but we that use or do not use them (Cramer, 2021). Clothing utilisation also has the advantage that it can be used for both individual garments and whole wardrobes, whether they are private, corporate, national or global. The goal we have to work towards is better utilisation overall, not that individual garments are used more at the expense of others. Possible sources for clothing utilisation rate complimenting consumer research are global production numbers, or import statistics to a specific country and the population statistics in the same area over time. This would demonstrate that there are planetary boundaries not only for production, but also for consumption. Reuse cannot increase the utilisation rate or the lifetime measured as the number of wears if there are not enough people compared to the amount of clothes. It is therefore impossible to increase clothing utilisation without decreasing production.

Conclusion

We have shown that the mode of acquisition as well as the volume of acquisition impacts the lifetimes of clothing both in terms of years and number of wears. At the same time, we have shown that there is a lot that we do not know about these relationships.

The link between the lifetime of clothes and the environmental impact is complicated. On one hand, it is not the case that lifetime necessarily affects what and how much is bought as we have discussed using the terms push and pull. On the other hand, a longer lifetime, or higher clothing utilisation will be necessary if clothing production is reduced. A link that we have not discussed, and where there is probably even less knowledge, is the one between how much new clothing is bought and how much is produced. To examine the impact of lifetimes on the environment, this connection must also be understood. There are many groups of clothing we have little knowledge about, and in particular, we have little knowledge about their use outside the global North and outside commercial circulation.

What most strongly affects the utilisation rate is the number of garments that are acquired. Extended service life for individual garments could mean that other clothes in the wardrobe are used less and thus do not necessarily improve the overall utilisation rate. To achieve a good utilisation rate measured as multiple uses per garment, the garments must be used over many years, and the number produced must be reduced.

The overall utilisation rate, whether we look at a private wardrobe, a country, or the world, is more important than the lifetime measured in years for an individual garment when discussing the environmental impact of clothing. Few uses per garment generate an increasingly less efficient industry, which thus develops in the opposite direction than the principles of the circular economy where the aim is to keep the products and materials in circulation at their highest use-value for as long as possible. Clothing lifetimes are affected by the number of items in the wardrobe, and therefore, more of the discussion should be about overall utilisation rather than measures, whether they are political or personal, that seek to increase the lifetime of individual garments.

As we have shown, the richness of forms of clothing acquisition practices affects the length of clothing lifespans. The prevailing discourse is that buying and selling used clothes is good, while we have shown that these items are worn fewer times, and the replacement rate is likely lower than previously estimated, though evaluating this accurately is complicated. Circular economy initiatives and business models should therefore prioritise the acquisition and use rather than discarding and reuse to start with, as this is where the greatest potential lies for keeping products in active use for a long time.

Consumers have a major impact on longevity and utilisation rate. Therefore, knowledge of consumption and not only design is important for understanding clothing lifetimes. As we have shown, a design-oriented understanding of what increases clothing lifetimes does not always concur with consumption research findings, such as the case of personalised garments. It is particularly important to investigate how different marketing strategies and business models affect consumption and thus also product lifetimes. This highlights the importance of evaluating design strategies and business models through empirical consumer research.

It is important to develop an understanding and relevant policy for clothing consumption that takes into account the complexity of the wardrobe to a greater extent. The wardrobes contain clothes with different use patterns; those that are used often, infrequently, clothes for occasions or specific activities, and so on. The findings therefore also have political implications. They can contribute to understanding which measures promote the longest garment lifetimes and utilisation rate with the least environmental impact and which kind of measures could help to implement these changes. This should be considered in life cycle assessments, as well as in policy development, where increasing the lifetime with the first user and slowing down the rate of acquisition should be the focus points. The work with increasing lifetimes should be combined with a reduction in acquisition so that the result is not larger wardrobes and longer lifetimes in years, instead, overall an increased utilisation of garments.

Acknowledgements

This work was supported by the Research Council of Norway, project nr. 303080 "LASTING-Sustainable prosperity through product durability", and Australian wool growers and the Australian government through Australian Wool Innovation Limited (AWI), contract number 4500012208. We would like to thank Nielsen AG for conducting the survey.

References

Barclaycard (2018). *Snap and Send Back* [online]. Available at: https://www. home.barclaycard/media-centre/press-releases/snap-and-send-back.html (Accessed: 2 May 2022)

Bernardes, J., Ferreira, F., Marques, A. and Nogueira, M. (2020). Consumers' clothing disposal behaviour: Where should we go? Textiles, Identity and Innovation. *In:* Touch: Proceedings of the 2nd International Textile Design Conference, Lisbon, June 19-21, 2019, CRC Press, doi: 10.1201/9780429286872-30.

Berthene, A. (2021) *Ecommerce is 46.0% of All Apparel Sales,* Digital Commerce 360 [online]. Available at: https://www.digitalcommerce360.com/article/online-apparel-sales-us/ (Accessed: 27 April 2022).

Binkley, C. (2012). The goal: A perfect first-time fit. *The Wall Street Journal*, March 23 [online]. Available at: https://www.wsj.com/articles/SB100014240527023047 24404577293593210807790 (Accessed: 31 May 2022).

Blázquez, M. (2014). Fashion shopping in multichannel retail: The role of technology in enhancing the customer experience. *International Journal of Electronic Commerce,* 18(4): 97-116. doi: 10.2753/JEC1086-4415180404

Bugge, A.B., Klepp, I.G., Borch, A., Schjøll, A., Laitala, K. and Haugrønning, V.A.R. (2019). *Jul - pynt, gaver, klaer og mat (Christmas – decorations, gifts, clothes and food),* Oslo: SIFO [online]. Available at: http://hdl.handle.net/20.500.12199/2928 (Accessed: 31 May 2022).

Cassidy, T.D. and Bennett, H.R. (2012). The rise of vintage fashion and the vintage consumer. *Fashion Practice: The Journal of Design, Creative Process & the Fashion,* 4(2): 239-262. doi: 10.2752/175693812X13403765252424

Castellani, V., Sala, S. and Mirabella, N. (2015). Beyond the throwaway society: A life cycle-based assessment of the environmental benefit of reuse. *Integrated Environmental Assessment and Management,* 11(3): 373-382. doi: 10.1002/ieam.1614

Citrin, A.V., Stem, D.E., Spangenberg, E.R. and Clark, M.J. (2003). Consumer need for tactile input: An internet retailing challenge. *Journal of Business Research,* 56(11): 915-922. doi: 10.1016/S0148-2963(01)00278-8

Connor-Crabb, A. and Rigby, E.D. (2019). Garment quality and sustainability: a user-based approach. *Fashion Practice,* 11(3): 346-374. doi: 10.1080/17569370.2019.1662223

Cooper, T. (ed.) (2010). *Longer Lasting Products: Alternatives to the Throwaway Society.* Surrey, UK: Gower Publishing Limited.

Corrigan, P. (1989). Gender and the gift: The case of the family clothing economy. *Sociology,* 23(4): 513-534. doi: 10.1177/0038038589023004002

Cramer, J. (2019). *The Living Wardrobe: Fashion Design for an Extended Garment Lifetime.* PhD, Melbourne, Australia: RMIT University.

Cramer, J. (2021). Use forecasting: Designing fashion garments for extended use. *In:* Muthu, S.S. and Gardetti, M.A. (Eds.), *Sustainable Design in Textiles and Fashion.* pp. 85-104. Singapore: Springer.

Cullinane, S., Browne, M., Karlsson, E. and Wang, Y. (2019). Retail clothing returns: a review of key issues. *In:* Wells, P. (Ed.), *Contemporary Operations and Logistics: Achieving Excellence in Turbulent Times.* pp. 301-322. Cham: Springer.

Cullinane, S. and Cullinane, K. (2021). The logistics of online clothing returns in Sweden and how to reduce its environmental impact. *Journal of Service Science and Management,* 14(1): 72-95. doi: 10.4236/jssm.2021.141006

Cullinane, S., Jansson, M., Browne, M. and Loon, P.V. (2021). *Consumer Attitudes towards Clothing Returns in Sweden: Preliminary Results of a Large-scale Survey.* Gothenburg: University of Gothenburg [online]. Available at: http://dx.doi.org/10.13140/RG.2.2.35038.15687 (Accessed: 31 May 2022).

Daily Mail (2015). Women ditch clothes they've worn just seven times: Items being left on the shelf because buyer feels they've put on weight or they've bought them on a whim. *Daily Mail.* 10 June [online]. Available at: https://www.dailymail.co.uk/femail/article-3117645/Women-ditch-clothes-ve-worn-just-seven-times-Items-left-shelf-buyer-feels-ve-weight-ve-bought-whim.html (Accessed: 31 May 2022).

Depop (2022). *The Power of Secondhand: How Resale Slows Consumption.* Depop and QSA partners, London. https://news.depop.com/download/docs/DepopDisplacementResearch2022.pdf.

de Wagenaar, D., Galama, J. and Sijtsema, S.J. (2022). Exploring worldwide wardrobes to support reuse in consumers' clothing systems. *Sustainability,* 14(1): 487. doi: 10.3390/su14010487

Domina, T. and Koch, K. (1997). The textile waste lifecycle. *Clothing and Textiles Research Journal,* 15(2): 96-102. doi: 10.1177/0887302x9701500204

Ellen MacArthur Foundation (2017). *A New Textiles Economy: Redesigning Fashion's Future.* Ellen MacArthur Foundation and Circular Fibres Initiative [online] Available at: https://ellenmacarthurfoundation.org/a-new-textiles-economy (Accessed: 28 September 2021).

Eurostat (2018). *Online Purchases.* EU28, 2018 [online] Available at: https://ec.europa.eu/eurostat/statistics-explained/index.php?title=File:Online_purchases,_EU28,_2018_(%25_of_individuals_who_bought_or_ordered_goods_or_services_over_the_internet_for_private_use_in_the_previous_12_months).png (Accessed: 27 April 2022).

Eurostat (2022a). E-banking and e-commerce – Internet use: Selling goods or services. [ISOC_BDE15CBC__custom_2759044] 30.03.2022 ed. Luxembourg [online]. Available at: https://ec.europa.eu/eurostat/databrowser/view/ISOC_BDE15CBC__custom_2759044/default/table?lang=en (Accessed 15 July 2022)

Eurostat (2022b). Internet purchases – Collaborative economy (2020 onwards): Online purchases (3 months) from private persons: any physical goods

[ISOC_EC_CE_I__custom_2759340], 30.03.2022 ed. Luxembourg [online]. Available at: https://ec.europa.eu/eurostat/databrowser/view/ISOC_EC_CE_I__custom_2759340/default/table?lang=en (Accessed 15 July 2022)

Farrant, L., Olsen, S. and Wangel, A. (2010). Environmental benefits from reusing clothes. *The International Journal of Life Cycle Assessment*, 15(7): 726-736. doi: 10.1007/s11367-010-0197-y

Fletcher, K. (2012). Durability, fashion, sustainability: The processes and practices of use. *Fashion Practice: The Journal of Design, Creative Process & the Fashion*, 4(2): 221-238. doi: 10.2752/175693812X13403765252389

Gabrielli, V., Baghi, I. and Codeluppi, V. (2013). Consumption practices of fast fashion products: A consumer-based approach. *Journal of Fashion Marketing and Management: An International Journal*, 17(2): 206-224. doi: 10.1108/JFMM-10-2011-0076

Gyde, C. and McNeill, L.S. (2021). Fashion rental: Smart business or ethical folly? *Sustainability*, 13(16): 8888. doi: 10.3390/su13168888

Hebrok, M., Klepp, I.G. and Turney, J. (2016). Wool you wear it? – Woollen garments in Norway and the United Kingdom. *Clothing Cultures*, 3(1): 67-84. doi: 10.1386/cc.3.1.67_1

Hebrok, M. and Heidenstrøm, N. (2017). *Maten vi kaster: En studie av årsaker til og tiltak mot matsvinn i norske husholdninger*, SIFO [online]. Available at: https://hdl.handle.net/20.500.12199/5337 (Accessed: 31 May 2022).

Henninger, C.E., Brydges, T., Iran, S. and Vladimirova, K. (2021). Collaborative fashion consumption – A synthesis and future research agenda. *Journal of Cleaner Production*, 319: 128648. doi: 10.1016/j.jclepro.2021.128648

Holroyd, A.T. (2014). Identity construction and the multiple meanings of homemade clothes in contemporary British culture. *In:* Hunt-Hurst, P. and Ramsamy-Iranah, S. (Eds.), *Fashion and its Multi-cultural Facets*. pp. 295-309. Leiden: Brill.

IMRG (2020). *IMRG Returns Review – 2020.* Interactive Media in Retail Group [online] Available at: https://www.imrg.org/insight/imrg-returns-review-2020/ (Accessed: 31 May 2022)

Johnson, E. (2020). *Dressing up the Environmental Potential for Product-service Systems: A Comparative Life Cycle Assessment on Consumption in Rental Clothing vs. Linear Business Models*. Master Thesis, Lund, Sweden: Lund University.

Jørgensen, M.S. and Jensen, C.L. (2012). The shaping of environmental impacts from Danish production and consumption of clothing. *Ecological Economics*, 83: 164-173. doi: 10.1016/j.ecolecon.2012.04.002

Joyner Armstrong, C.M. and Park, H. (2020). Online clothing resale: A practice theory approach to evaluate sustainable consumption gains. *Journal of Sustainability Research*, 2(2): e200017. doi: 10.20900/jsr20200017

Klepp, I.G., Laitala, K., Schragger, M., Follér, A., Paulander, E., Tobiasson, T.S., Eder-Hansen, J., Palm, D., Elander, M., Rydberg, T., Watson, D. and Kiørboe, N. (2015). *Mapping Sustainable Textile Initiatives and a Potential Roadmap for a Nordic Actionplan*, Nordic Council of Ministers [online]. Available at: http://norden.diva-portal.org/smash/get/diva2:840812/FULLTEXT01.pdf (Accessed: 31 May 2022).

Klepp, I.G. and Laitala, K. (2016a). *Klesforbruk i Norge*, Oslo: SIFO [online]. Available at: https://hdl.handle.net/20.500.12199/5331 (Accessed: 31 May 2022).

Klepp, I.G. and Laitala, K. (2016b). "Ullne" fakta om strikking og klær. Hjemmeproduksjon og gamle klær i velstands-Norge. *In:* Lavik, R. and Borgeraas, E. (Eds.), *Forbrukstrender 2016*. Oslo: SIFO. Available at: https://hdl.handle.net/20.500.12199/2980 (Accessed: 14 July 2022)

Klepp, I.G. and Laitala, K. (2018). Shared use and owning of clothes: Borrow, steal, or inherit. *In:* Cruz, I.S., Ganga, R. and Wahlen, S. (Eds.), *Contemporary Collaborative Consumption – Trust and Reciprocity Revisited*. Wiesbaden, Germany: Springer VS.

Klepp, I.G., Laitala, K. and Haugrønning, V. (2019). Wardrobe sizes and clothing lifespans. *In:* Jaeger-Erben, M. and Nissen, N.F. (Eds.), *Plate*, 18-20 September 2019 Berlin. Universitätsverlag der TU Berlin, 451-456 [online]. Available at: https://depositonce.tu-berlin.de/handle/11303/10291 (Accessed: 31 May 2022).

Klepp, I.G., Laitala, K. and Wiedemann, S. (2020). Clothing lifespans: What should be measured and how. *Sustainability*, 12(6219): 21. doi: 10.3390/su12156219

Klepp, I.G. and Tobiasson, T. (2021). *Lettfiks. Klær med ni liv*, Oslo: Bokvennen Forlag.

Kumar, V., Ekwall, D. and Zhang, D.S. (2022). Investigation of rental business model for collaborative consumption – workwear garment renting in business-to-business scenario. *Resources, Conservation and Recycling*, 182: 106314. doi: 10.1016/j.resconrec.2022.106314

Lai, C.-C. and Chang, C.-E. (2020). Clothing disposal behavior of Taiwanese consumers with respect to environmental protection and sustainability. *Sustainability*, 12(22): 9445. doi: 10.3390/su12229445

Laitala, K. and Boks, C. (2012). Sustainable clothing design: Use matters. *Journal of Design Research*, 10(1/2): 121-139. doi: 10.1504/JDR.2012.046142

Laitala, K. (2014). Consumers' clothing disposal behaviour – A synthesis of research results. *International Journal of Consumer Studies*, 38(5): 444-457. doi: 10.1111/ijcs.12088

Laitala, K. and Klepp, I.G. (2015). Clothing disposal habits and consequences for life cycle assessment (LCA). *In:* Muthu, S.S. (Ed.), *Handbook of Life Cycle Assessment (LCA) of Textiles and Clothing*. Cambridge: Elsevier.

Laitala, K. and Klepp, I.G. (2017). Clothing reuse: The potential in informal exchange. *Clothing Cultures*, 4(1): 61-77. doi: 10.1386/cc.4.1.61_1

Laitala, K. and Klepp, I.G. (2018). Motivations for and against second-hand clothing acquisition. *Clothing cultures*, 5(2): 247-262. doi: 10.1386/cc.5.2.247_1

Laitala, K. and Klepp, I.G. (2020a). *Klær og miljø: Innkjøp, gjenbruk og vask*, Oslo: SIFO [online]. Available at: https://hdl.handle.net/20.500.12199/3065 (Accessed: 28 September 2021).

Laitala, K. and Klepp, I.G. (2020b). What affects garment lifespans? International clothing practices based on wardrobe survey in China, Germany, Japan, the UK and the USA. *Sustainability*, 12(21): 9151. doi: 10.3390/su12219151

Laitala, K. and Klepp, I.G. (2021). Clothing longevity: The relationship between the number of users, how long and how many times garments are used. 4th PLATE Virtual Conference. Limerick, Ireland, 26-28 May 2021. Available at: http://hdl.handle.net/10344/10223 (Accessed: 15 July 2022).

Lambert, M. (2004). "Cast-off Wearing Apparell": The consumption and distribution of second-hand clothing in northern England during the long eighteenth century. *Textile History*, 35(1): 1-26. doi: 10.1179/004049604225015620

Lang, C. and Zhang, R. (2019). Second-hand clothing acquisition: The motivations and barriers to clothing swaps for Chinese consumers. *Sustainable Production and Consumption,* 18: 156-164. doi: 10.1016/j.spc.2019.02.002

Lee, S.H. and Cotte, J. (2009). Post-purchase consumer regret: Conceptualization and development of the PPCR scale. *In:* McGill, A.L. and Shavitt, S. (Eds.), *NA – Advances in Consumer Research.* 36, pp. 456-462. Duluth, MN: Association for Consumer Research.

Lehmann, M., Tärneberg, S., Tochtermann, T., Chalmer, C., Eder-Hansen, J., Seara, J.F., Boger, S., Hase, C., Berlepsch, V.V. and Deichmann, S. (2018). *Pulse of the Fashion Industry.* Global Fashion Agenda & The Boston Consulting Group [online] Available at: http://www.globalfashionagenda.com/download/3700/ (Accessed: 31 May 2022).

Levänen, J., Uusitalo, V., Härri, A., Kareinen, E. and Linnanen, L. (2021). Innovative recycling or extended use? Comparing the global warming potential of different ownership and end-of-life scenarios for textiles. *Environmental Research Letters,* 16(5): 054069. doi: 10.1088/1748-9326/abfac3

Maldini, I. and Balkenende, A.R. (2017). Reducing clothing production volumes by design: A critical review of sustainable fashion strategies. *In:* Bakker, C. and Mugge, R. (Eds.), *Product Lifetimes and the Environment - PLATE 2017,* pp. 233-237. 9 November 2017. Delft: Delft University of Technology and IOS Press [online]. Available at: http://ebooks.iospress.nl/publication/47876 (Accessed: 15 July 2022).

Maldini, I., Duncker, L., Bregman, L., Piltz, G., Duscha, L., Cunningham, G., Vooges, M., Grevinga, T., Tap, R. and Balgooi, F.V. (2017). *Measuring the Dutch Clothing Mountain. Data for Sustainability-oriented Studies and Actions in the Apparel Sector.* Amsterdam University of Applied Sciences [online] Available at: https://research.hva.nl/files/3144178/Measuring_the_Dutch_Clothing_Mountain_final_report_002_.pdf (Accessed: 14 July 2022).

Maldini, I., Stappers, P.J., Gimeno-Martinez, J.C. and Daanen, H.A.M. (2019). Assessing the impact of design strategies on clothing lifetimes, usage and volumes: The case of product personalisation. *Journal of Cleaner Production,* 210: 1414-1424. doi: 10.1016/j.jclepro.2018.11.056

Mauss, M. (1970). *The Gift: Forms and Functions of Exchange in Archaic Societies.* London: Cohen & West.

McAfee, A., Dessain, V. and Sjöman, A. (2004). *Zara: IT for Fast Fashion,* Boston, MA: Harvard Business School Publishing.

Miller, D. (1998). *A Theory of Shopping.* Cambridge: Polity Press.

Morgan, L.R. and Birtwistle, G. (2009). An investigation of young fashion consumers' disposal habits. *International Journal of Consumer Studies,* 33(2): 190-198. doi: 10.1111/j.1470-6431.2009.00756.x

Nielsen, Ag (2019). *Global Wardrobe Audit* (dataset). For Australian Wool Innovation Ltd. Sydney, Australia: The Nielsen Company.

Niinimäki, K. and Koskinen, I. (2011). I love this dress, it makes me feel beautiful! Empathic knowledge in sustainable design. *Design Journal,* 14(2): 165-186. doi: 10.2752/175630611x12984592779962

Nørup, N., Pihl, K., Damgaard, A. and Scheutz, C. (2019). Replacement rates for second-hand clothing and household textiles – A survey study from Malawi, Mozambique and Angola. *Journal of Cleaner Production,* 235: 1026-1036. doi: 10.1016/j.jclepro.2019.06.177

Parker, B. and Weber, R. (2013). Second-hand spaces: Restructuring retail geographies in an era of E-commerce. *Urban Geography,* 34(8): 1096-1118. doi: 10.1080/02723638.2013.790642

Rathinamoorthy, R. (2020) Clothing disposal and sustainability. *In:* Muthu, S. and Gardetti, M. (Eds.), *Sustainability in the Textile and Apparel Industries.* pp. 89-120. Cham: Springer.

Reed, C. (2019). *Launching the Australasian Circular Textile Association.* The Australasian Circular Textile Association (ACTA) [online]. Available at: https://www.australiancircularfashion.com.au/launching-acta/ (Accessed: 24 May 2022).

Roichman, R., Makov, T., Sprecher, B., Groot, L.D., Blass, V. and Shabtai, S. (2023). The untold story of e-commerce product returns and its life cycle environmental impacts. *In:* Niinimäki, K. (ed.), *Plate – Product Lifetimes and the Environment.* Espoo: Aalto University.

Shim, S. (1995). Environmentalism and consumers' clothing disposal patterns: An exploratory study. *Clothing and Textiles Research journal,* 13(1): 38-48. doi: 10.1177/0887302X9501300105

Statista (2019). *Share of internet users who have purchased selected products online in the past 12 months as of 2018* [online] Available at: https://www.statista. com/statistics/276846/reach-of-top-online-retail-categories-worldwide/ (Accessed: 27 April 2022).

Statista (2022). *E-commerce as percentage of total apparel sales in Europe from 2018 to 2025,* Statista [online] Available at: https://www.statista.com/statistics/1277311/ e-commerce-share-apparel-sales-europe/ (Accessed: 2 May 2022).

Stevenson, A. and Gmitrowicz, E. (2012). *Study into Consumer Second-hand Shopping Behaviour to Identify the Re-use Displacement Effect.* WRAP.

Strathern, M. (2011). Sharing, stealing and borrowing simultaneously. *In:* Strang, V. and Busse, M. (Eds.), *Ownership and Appropriation.* Oxford: Academic Press.

Tinson, J. and Nuttall, P. (2007). Insider trading? Exploring familial intra-generational borrowing and sharing. *The Marketing Review,* 7(2): 185-200. doi: 10.1362/146934707X198885

Vadenbo, C., Hellweg, S. and Astrup, T.F. (2017). Let's be clear(er) about substitution: A reporting framework to account for product displacement in life cycle assessment. *Journal of Industrial Ecology,* 21(5): 1078-1089. doi: 10.1111/jiec.12519

van Nes, N. and Cramer, J. (2006). Product lifetime optimization: A challenging strategy towards more sustainable consumption patterns. *Journal of Cleaner Production,* 14(15-16): 1307-1318, doi: 10.1016/j.jclepro.2005.04.006

Vesterinen, E. and Syrjälä, H. (2022). Sustainable anti-consumption of clothing: A systematic literature review. *Cleaner and Responsible Consumption,* 5: 100061. doi: 10.1016/j.clrc.2022.100061

Part III

Recycling

Waste Not, Want Not: Re-routing Garment and Textile Waste Streams to Create Circular Fashion Design Eco-systems in Bangladesh and Beyond

Anne Peirson-Smith[1]* and Stella Claxton[2]

[1] Northumbria University
[2] Nottingham Trent University
e-mail: *anne.peirson-smith@northumbria.ac.uk

Collective concern about unrestrained clothing production and consumption

There is widespread concern about the scale of the environmental and social impacts caused by the continued and unrestrained growth in global textile and clothing production and consumption. Some sources suggest that by 2030, the global consumption of garments might rise to 63% from 62 million tons to 102 million tons, with the retail value of garments and footwear increasing by 30% to 2 trillion Euros, representing an increase of 500 billion Euros (Global Fashion Agenda, 2017). Awareness of this extant environmental crisis facing the planet has necessitated a call to action by consumer and producer groups involved in the debate on sustainable fashion in which governments, policy makers, industry and academic researchers are increasingly seeking solutions within a design innovative eco-system (Whicher and Walters, 2017) based on nine factors: design users (private sector, public sector and general public); design support; design promotion; design agents (centres, associations, networks and clusters); professional design sector; design education; research and knowledge exchange; funding; policy, governance and regulation (Whicher, 2017, p. 121). Unlike previous fragmented multi-stakeholder initiatives, this integrated eco-system approach recognises the strategic potential of design in locating innovative solutions. In this case it can be framed within a circular economy, for example, by creating new products and services, rethinking organisational management and production and tapping into market and consumer trends, resulting in beneficial sustainable societal and economic outcomes (Na et al., 2018).

Representative signatories across a range of communities of practice have focussed on generating sustainable solutions for the fashion and textile sectors. This discourse forms part of a broader, long-standing debate on potential sustainable solutions to scale and manage resources used in textile and garment production and consumption, aligning with the key findings of climate science (McDonough and Braungart, 2002). The parties involved attempt to address systemic solutions intended to stem the tide of climate change and catastrophic planetary impact by aiming to reduce greenhouse gas (GHG) emissions by at least 25% and to restrict global warming at 1.5 degree centigrade, pre-industrial levels as sanctioned by the 2016 UNFCCC Paris Climate Change Agreement (UN, 2015). This consensus of concern across the design eco-system (Whicher, 2017) is gaining traction, suggesting that the textile and garment industry should replace the root of the issue – the negative environmental impact of its current extractive, take, make, waste system (Ellen MacArthur Foundation, 2013; Global Fashion Agenda, 2017) with a circular solution. This could be based on innovative design thinking and strategic operations management that supports an industrial economy guided by regenerative and restorative values and design principles. A circular, or closed loop approach intends to minimise waste and the dependency on virgin raw materials by recycling post-industrial and post-consumer waste materials into new products (McDonough and Braungart, 2013; Vecchi, 2020, Moreira and Niinimäki, 2022; Coste-Maniere et al., 2019) as a way of minimising preventable environmental impact to protect people and preserve natural resources, while generating value for incumbent stakeholders across the global design eco-system. Campaigners for the circular economy propose that operationalising embedded circular business models supports the move to a more sustainable, decarbonised global fashion and textiles industry by reducing pollution and waste, keeping products in use for longer, and transitioning to renewable energy sources and materials.

To effect impactful changes in current industrial practices, targeted systemic change needs to be accelerated by moving away from piecemeal sustainability initiatives focussing on vague commitments to become carbon neutral or use of recycled materials to achieve more resilient, integrated design-led solutions that encompass the entire fashion system (Palm et al., 2021). This can be directed by multi-stakeholder representation acting from the upstream and downstream supply chains, backed up by tangible action and driven by investment in new technological systems, incentivised by legislation. Consequently, the transformative agenda for the textile and garment industry requires the involvement of many actors and stakeholders working together in a pre-competitive space (Mishra et al., 2021; Brydges, 2021) to address the environmental impact of an extractive textile and garment industry reliant on 98 million tonnes of non-renewable and virgin resources per annum (Ellen MacArthur Foundation, 2017).

The recognition of the need for collaborative action galvanised by innovative design eco-system thinking reflects an urgency to shift to a climate-neutral, circular textile and fashion economy with all products designed for longevity and circularity, being embedded with durability, reusability, repairability, recyclability and produced with reduced environmental impact. In practice, the scaling of a circular approach requires applying design standards, lengthening product and material lifetimes and usage, while assuring that parts of the modular whole of a given product can be dismantled and repurposed or recycled into new textile and clothing products, or used for application in other industries (Cooper and Claxton, 2022). Furthermore, circularity requires improved transparency of supply chain operations to allow accurate tracking of materials and environmental impacts, along with the development and scaling of textile and garment collection and recycling technology and infrastructure.

The collated wisdom from collaborative initiatives within public and private sector initiatives, such as the WRAP Sustainable Clothing Action Plan, and through membership of industry bodies and knowledge exchange platforms such as UK based professional network, Fashion and Textile Association (UKFT) and ethical networking hub, Common Objective has identified four key areas of concerted action that are required from the industry in terms of (i) minimising upstream operational emissions; (ii) increasing closed loop recycling initiatives across the supply chain; (iii) reducing operational emissions and (iv) enabling more sustainable consumer behaviour (Ellen MacArthur Foundation, 2017; Bukhari et al., 2018). While this fourfold action plan is vital to affect any sort of holistic regenerative change across the textile and garment supply and value chains, the focus on recalibrating upstream manufacturing operations and recycling post-industrial waste has been relatively overlooked in favour of investigating post-use, post-consumer waste initiatives based on implementing collection and recycling systems in addition to producing durable textiles and designing clothing for longevity, involving both producers and consumers (Norris, 2019; Kozlowski et al., 2012). Grounded brand and consumer-oriented solutions for post-consumer waste are critical for circularity, given the larger volume of resulting waste and the ten-fold increase in clothing consumption and production over the past decade (Niinimäki et al., 2020), exacerbated by increased spending power in the Global North, in addition to worldwide population growth and global industrial expansion and manufacturing plants in the Global

South. However, more attention also needs to be paid to the potential of upstream textile waste to minimise its environmental impact and recoup its value based on systemic circular solutions. As the Reverse Resources report noted "garment manufacturers producing textiles and clothes for many of the world's major fashion brands are spilling an average of 25% of leftover volumes out of production…the volume is as high as 47%, much higher than usually perceived by brands" (Runnel et al., 2017).

Despite the growing recognition of the urgent need to find workable solutions, the textile and clothing industry response to the speed and scale of change needed to meet national/global environmental and carbon targets appear to be insufficient and is seemingly on course to "overshoot its 2030 carbon reduction targets by almost twofold by 2030 unless action is significantly accelerated" (McKinsey, 2020), highlighting the vital need to locate and scale-up all abatement initiatives. Equally, there is still a sense (Global Fashion Agenda, 2021) that the textile and fashion industry and its stakeholders, so far, have only been guided by non-binding voluntary agreements and self-regulation which has yielded piecemeal action to date. Manufacturing operations in emerging countries in the Global South also lack infrastructure, government support or connectivity. Other commentators even question the validity and viability of circular systems application (Earley and Goldsworthy, 2018; Garcia-Torres et al., 2017) as the total sustainability solution for communities of practice involved in the garment and textile industry (see Corvellec et al., 2022) for example, by questioning the circularity credentials of certain fibres such as recycled polyester made from rPET plastic bottle waste, which is not in itself recyclable once in textile form.

Policy and legislative incentives for sustainability action

Nevertheless, the industry, policy makers and governmental representatives have largely recognised that this systemic problem needs to be fixed, and that tangible progress has been made among collaborative industry initiatives, such as the Waste and Resources Action Programme (WRAP), Textiles 2030 (WRAP, 2022) the EU Strategy for Sustainable and Circular Textiles (European Commission, 2022) and British Fashion Council's Circular Fashion Ecosystem Report (BFC, 2021). To exemplify, WRAP's Sustainable Clothing Action Plan (SCAP) has enabled the textile and clothing industries to adopt evidence base and tools to meet targets reducing their carbon footprint across 8 years by 21.6%; the water footprint by 18.2%; the waste footprint by 2.1% and clothing waste by 4%, a collective industry effort that continues with Textiles 2030 (WRAP, 2022). Globally, regulators are activating similar circularity concepts with China's 2021 five-year plan to launch a circular economy by supporting recycling, re-manufacturing and the preferred use of renewable energy, for example (Li et al., 2021). The EU's Green Deal directive, as part of its Circular Economy Action Plan 2021 (Friant et al., 2021) proposes to make all goods throughout their lifetime more environmentally friendly and circular and energy efficient, in addition to singling out textiles to make them more reusable, recyclable and durable to minimise waste. Equally, the EU's Waste Directive Framework expects that by 2025 all member countries will separate and reuse textile waste. Some European nations have already implemented extended producer responsibility schemes requiring brands and retailers to take ownership of post-consumer waste and expecting producers to make financial contributions for the collection, recycling and reuse of their products. In response to the unhindered disposal of pre- and post-consumer textile and garment waste in Europe via incinerators (30%) and landfill (70%), the latter costing €60 per ton, the EU has implemented new sustainable waste management rules centring on incentivising closed-loop recycling processes that reduce and recycle materials into biodegradable waste. The extension of the Extended Producer Responsibility (EPR) regulatory instrument from 2023 to textile production for member states across the EU, and potentially in the UK, will require producers to assume financial and material responsibility for the handling of marketable products to encourage the prevention of waste generation at source. This requires the implementation of reduced impact product design, and incentivising wider public re-use and recycling of garments by accelerating post-consumer take back, recovery and disposal schemes (WRAP, 2022; WRAP, 2013), aligned with reuse and recycling targets to further develop an economy that supports sustainability.

In particular, the EU Strategy for Sustainable and Circular Textiles (European Commission, 2022) outlines a vision for a resilient and innovative circular textiles ecosystem that by 2030 textile products in the EU marketplace are recyclable and have longevity, comprise of recycled fibres, eliminate hazardous substances, manage waste, and eliminate incineration and disposal, while being manufactured in line with human rights and environmental responsibility overall. To operationalise this twin agenda, driven by sustainability and technological goals, the EU directive recommends better information management, based on digital passport systems using QR codes or RFID and smart label technology along with the extended producer responsibility scheme. Aiming to kickstart this transformative journey, a co-creative initiative, the "transition pathway for the textiles ecosystem" is underway. This intends to incorporate feedback from various players in the European textile and garment system (European Commission, 2022) to demonstrate and share their commitments on circularity and circular business models, highlight actions to strengthen sustainable competitiveness, digitalisation and resilience, and identification of specific investments needed for this dual transition (European Environmental Agency, 2022).

There are promising signs from a regulatory perspective incentivising closed-loop systems driven in part by urging innovation and collaboration to obviate some of the barriers to the recycling of waste based on collection and sorting. These policy and legislative initiatives propose moves towards design practices using circular, closed loop business models that facilitate extended sharing, leasing, reusing, repairing, refurbishing and recycling of materials and products, sustainability metrics to verifying progress toward cleaner and more sustainable production and management of water and waste, although not without controversy. For example, Sustainable Apparel Coalition's (SAC) recently contested synthetics-biased Higg Index (Bridge, 2022) perhaps emphasises the need to devise and implement more reliable national/global environmental policy and legislation to necessitate viable change by making the supply chain more transparent and connected (Hole and Hole, 2020). This understanding is based on investment in enabling technology that can track and trace the lifecycle of textiles in terms of their origins and environmental impact. Transparent and meaningful data on how textiles are processed into garments at each step of the supply chain should be accessible to all stakeholders involved in production and consumption (Garcia-Torres et al., 2021).

Innovative technical solutions to industrially scale circularity

In response to pending legislation requiring producers to recognise and visibly demonstrate more environmental responsibility for their production practices and designed products, significant investment in portfolios of innovative technology and digitisation using Blockchain technology for example, to access and share vital production and product information such as adopting product passports or devising smart integrated supply chains based on open source cloud computing is taking place on the part of textile and garments suppliers and brands (Kayikci et al., 2022). Yet, pressure is mounting on the industry to scale-up regenerative initiatives ideally aiming to eliminate textile waste while keeping resources in perpetual circulation, such as *Worn Again's* chemical textile recycling technology that can separate blended fibre combinations.

Regenerative recycling technology would enable circularity by allowing fibres to go back into manufacturing, replacing virgin materials and providing long-term sustainable resource security. Mechanical recycling of mono-fibre wool and cotton textiles is already well-established and has the potential to be further scaled up. However, advances in regenerative recycling technologies for synthetic and man-made cellulosic fibres (MMCF), and mixed fibres are limited either by high cost and lack of capacity (pointing to a need for investment) or by inferior fibre quality (GFA and McKinsey, 2021). Technological innovations in the fashion system appear to have reached a tipping point and are evolving from pilot to industrialised proofs of concept (BoF, 2022). The pressure to implement transparency and demonstrate accountability across the supply chain and the activation of circular business models will further push for more widespread application and coordination. These trends are exemplified by the Connected Product Platform by Eon enabling client brands, such as Yoox Net-a-Porter to curate information about the full lifecycle provenance of their products, while Blockchain driven platform FibreTrace is partnering with Reformation allows consumer QR coded access to its denim product lifecycle and IBM and Red Hat have collaborated on a smart supply chain system for transparent data management and logistics (Business of Fashion, 2021; Gutiérrez, 2018).

Difficulties inevitably exist in implementing producer responsibilities for recycling and re-using post-consumer garment waste. This is down to the scale of the problem and the lack of infrastructure to create a system to accommodate this from improving time-consuming manual collection and automating the sorting systems of varied fibre blends with various trims, to managing resulting textile waste by investing in reuse and recycling technologies such as anaerobic digestion, fermentation, composting, regenerating fibres and thermal recovery systems. More research appears necessary to identify "the structure and properties of cellulosic fibres regenerated from cotton-based waste" and investigate "recycling technologies to sustainably manage other textile waste, such as man-made cellulosic fibre (MMCF) and other fibres (polyamide, wool, rayon, silk, acrylic)" (Juanga-Labayen et al., 2022, p. 184) and these technological approaches should be implemented holistically.

Re-focus on post-industrial textile waste

The focal point of this chapter examines both the management and destination of industrial textile waste. It examines post-industrial waste created from fibres, yarns and sortable clean textiles during the garment manufacturing process, often in 'mint' condition. Offcuts from the cutting process in a production run, for example, represent clean and sortable textiles on account of their homogenous colour and fibre composition that can be reclaimed into yarns or upcycled into other garments without too much complication (Domina and Koch, 1997). This contrasts to post-consumer waste whereby clothing is devalued and discarded after service due to damage from washing, wearing, and staining (Hvaas and Pedersen, 2019; Binotto and Payne, 2017) that is compounded by the challenges of sorting, classifying, and recycling. Focussing on repurposing post-industrial textile waste offers a concrete material circular solution that appears to be achievable and scalable (Aus et al., 2021). Some sources even consider untapped post-industrial textile waste streams to be a "low hanging fruit that we can't afford to miss", that are technically "more consistent, higher quality, easier to trace, significantly larger than previously estimated and require much simpler sorting and collection than post-consumer waste" (Global Fashion Agenda, 2021). Ideally, given the high potential for post-industrial textile and garment recycling, most could be reused and recovered, with only a small percentage ending up being discarded or diverted.

While the recycling of textile waste is only one part of the circularity solution, it is a significant one in terms of its potential to attain sustainability objectives and fulfil environmental and economic goals via industrial operations. It also has future potential, given that globally only 1% of textile waste is being currently recycled into renewed fibres for garment production, with 25% reused or recycled in general, and around 75% ending up in landfills (McKinsey, 2022). Overall, there is little evidence that these operations are being significantly scaled-up in global terms, despite the material cost comprising 2% of each garment. In reality, the majority of post-industrial textile waste is discarded into landfills or incinerated.

The scale of the unmitigated post-industrial textile waste problem is evident. From a global volume production of textile fibres totaling 111 million metric tonnes (MMT) in 2019, the overall production of the synthetic textile, polyester, alone represents 51% of all textiles produced at 54 MMT per year, with cotton coming in second, constituting 25% or 26 MMT per annum (Juanga-Labayen et al., 2022). Alongside the business case and economic potential of circularity, the need to effect systemic change in industrial textile and garment production is also pertinent from an environmental angle, given that material production is the biggest culprit regarding negative impact in terms of the magnitude of water usage and pollution, notwithstanding consequent fertiliser and chemical impact on land, soil and communities in the production country of origin, for example. Of course, recycling and reuse processing using mechanical, chemical, and biological options for recycling textile waste into products involves energy and raw material consumption, while creating emission and environmental impact on soil, water, and air. Arguably, this impact can be minimised operationally.

The upstream waste management opportunity

Work on recycling textile and apparel waste has highlighted various means of recovering, reusing or recycling material textile waste in the form of fibre, yarn, and fabric scraps or larger quantities of rejected waste fabric by ensuring their biodegradability pre-production qualities, enabling them to be repurposed into new raw materials (Wang, 2010; Ekström and Salomonson, 2014). Other studies have focussed on the management of rejected, unsold or discarded post-consumer clothing waste, often extending beyond the site of production and country of origin to other geographic regions such as wholesale markets for waste garments, aiming to keep materials in the loop for at least another circuit (Domina and Koch, 1997; Brooks, 2013; Bukhari et al., 2018).

It has been suggested that with additional focus and investment that over 50% of spinning waste output from textile and garment mills and factories could be recycled into new fibres for new garments (Textile Exchange, 2021). In addition, 25% of overstock and deadstock fabrics, in another circuit of production, could be adapted and reused in-house, thereby optimising the production process (Bukhari et al., 2018). The re-routing of 'spent' textiles from landfill or incineration lowers the textile industry's GHG emissions, while simultaneously reducing the utilisation of virgin materials, thereby minimising environmental impact.

A report by textile tracking and trading company, Reverse Resources, (Reverse Resources, 2021) determined the high percentage of waste that exists for recycling in the global market based on fibre compositions. The company examined the fibre composition breakdown referenced in the Preferred Fiber and Materials Market Report by the Textiles Exchange (Textiles Exchange, 2020) and plotted this data against their own prior waste mapping surveys across 20 countries involving over 1200 factories. According to their analysis, Reverse Resources estimated that industrial recyclable textile waste has a potential total volume of 9 million metric tonnes (MMT) from 37 MMT of fibre produced annually by the industry. This breakdown comprises most polyester-based blends at 3.62 MMT, followed by 100% polyester at 1.21 MMT, and cotton or cotton blends at 1.08 mln metric tonnes each, in addition to polyamide at 0.47, other synthetics at 0.54, MMCFs at 0.59, other plant based at 0.54 and the remaining share by wool, silk and down. Typically, this reflects the main synthetic fibres, polyester, acrylic and nylon used to manufacture garments globally, followed by their naturally sourced equivalents of cotton, wool, and flax, for example. Normally, waste produced during the manufacture of textiles and garments results is estimated to range from 12%-30% (Runnel et al., 2017) of the original material produced comprising fibre, yarn, fabric, and clothing items. This includes a broad inventory of cotton lint, sub-standard or damaged yarns as a by-product of yarn production and spinning used as feedstock for composites; fly fibres and greige or unfinished fabrics in knitted and woven fabrics, poor quality dyed and finished fabrics to excess fabrics, in addition to the cut pieces and samples or over-produced garments remaining following various garment manufacturing stages (Juanga-Labayen et al., 2022). In terms of the upcycling potential for this discarded fabric, "depending on the size of the factory the textile waste and fabric leftovers generated in the garment manufacturing process ranges from 25 to 40% of the total fabric used...50%

of that material can be upcycled into new garments and for some types of leftover...it can even be up to 80%" (Aus et al., 2021, pp. 15-16).

Solutions to post-industrial waste recycling and circularity

Seemingly, there is both the will and the way to implement circular systems in general, and a desire to scale-up the main reliance on virgin materials by commercially ramping-up waste fibre and yarn recycling of natural or synthetic polymers using primary recycling technologies. Technological solutions can facilitate the recycling of post-industrial textiles and garment waste in the form of cutting scraps, sub-standard pieces, overproduction, leftovers and offcuts to redirect spent textiles from landfill or incineration.

Given its recycling potential and embedded value, post-industrial waste should not necessarily be downgraded and labelled negatively as 'waste', but rather can be upgraded as a key re-usable resource, offering viable and identified solutions, thereby creating a raw material source stream and contributing to closing the loop for textile and garment manufacturing. Textile waste recycling typically relies on mechanical or chemical processing. The former mechanically recycled option, whether cotton or wool based, is highlighted by the industry benchmark espoused by the SAC's HIGG Index, and the output is considered to yield a lower GHG carbon footprint of up to 70%, as compared to its conventional virgin material equivalents, such as fossil-fuel formed polyester.

A "closed-loop" recycling initiative (CLR), rather than an "open loop" system, is geared to recoup raw materials used in the manufacturing of a polymer product, by reprocessing it into a similar product of equivalent quality as its virgin material origin. As Payne notes, "CLR, collected textiles are reprocessed into new fibre to use for new garments, re-entering the same production system that the textile originally came from. In more sophisticated approaches to CLR, the recycled fibre can be repeatedly recycled", yet may not retain the same quality (Payne, 2015, p. 106).

To convert post-industrial waste into a valued commodity, processing and sorting by fibre content or colourway, for example, is required, ideally at factory level. Equally, this is often more straightforward, because post-industrial waste is relatively standardised regarding colour, chemical components, and composition, which also pre-disposes this type of waste to recycling interventions. To facilitate these initiatives, the technologies already exist "to deliver recycling across single colour cotton, cellulosics, synthetic fabrics and solutions for blended fibres" (GFA and McKinsey, 2021), and are evolving to enable recycling as a viable, scaled commercial option with a reduced environmental impact. The main issue is having access to these technological resources. According to McKinsey's modelling efforts, existing technology can support 75% textile-to-textile recycling back into the production system, supplemented by an additional 5% feedstock from other industrial sources. However, these initiatives are still considered as a work in progress urgently needing investment in and scaling-up in developing countries, where the textile and garment manufacturing plants are largely located on account of the availability of low-cost labour and existing skills.

Challenges of textile and garment waste recycling scale-up

Manufacturers aim to maximise financial value from textile waste. It is estimated that the critical challenges reside in providing conditions for scaling circularity initiatives, which in this case include the collection and sorting infrastructure, and technological investment in the recycling sector to scale-up capacity (McKinsey, 2022). Nevertheless, suppliers need to work closely with brands to optimise their production processes through lean manufacturing methods and careful management of output, given that "a factory in Dhaka…simply by rearranging how fabrics are fixed on the cutting table…managed to reduce the length of each fabric lay by 3 cm and achieved 22,000 USD savings per month" (Runnel et al., 2017).

Despite efficient manufacturing methods and product optimisation by manufacturers relating to cut and sew processes (Townsend and Mills, 2013), the unofficial waste management system in manufacturing countries typically in the Global South, is informal and the availability of different recycling or upcycling options is often limited relating to fabric quality or manufacturing and resource planning (Aus et al., 2021). Production volume data on leftover textile material is reputedly inaccurate and under-reported (Runnel et al., 2017). It is estimated that 25% of overspills from garment manufacturing exit the loop and end their life being incinerated or discarded in landfill (Leonas, 2017), while less than 1% of textile waste is fibre-to-fibre recycled. Beyond this, the deliberate lack of accounting occurs on the part of suppliers as brands already assume that they have paid in their production orders up to 10% to cover production inefficiency including textile scraps, errors and over-runs. In actuality, as knits are bought by weight in kilograms and ready-made garments (RMG) by number of pieces there is an information gap in tracking the ratio of input to output. Given that "the current dominant method of production costing gives a monetary incentive for garment suppliers to underreport leftover information after production is finished" (Runnel et al., 2017) suppliers can get some return on unaccounted for overspill, making up to 25% of their profit margin, yet the majority of this overspill is dumped and ends its life.

Equally, the production facilities for textiles and garments are often located in developing Asian countries where the investment in technological development is patchy and quality control varies with production practices being non-transparent and equitable labour practices being systematically transgressed by factory owners (Ross, 2004). Studies have highlighted the barriers to implementing scaled fibre and textile recycling ventures being rooted in accessibility and economics (Reinecke et al., 2019). The cost of implementing physical and sorting operations and investment in technological processing systems, locating a workforce or identifying a reliable market for the resulting recycled output, all constitute barriers to actioning textile recycling initiatives at ground level.

The relative neglect regarding the use of post-industrial waste to date is accounted for in terms of systemic barriers (Leal Filho et al., 2019). This oversight in recognising potential value which has been estimated at 18-26% of fibre-to-fibre rate of recycling by 2030 in Europe alone with 70% of textile waste having the potential to be recycled fibre-to-fibre (McKinsey, 2022), may also be partly due to embedded operational power relations on the supply side, and the relative under-representation of suppliers in collaborative think tanks and discussion platforms. The RMG industry in Bangladesh, for example, is frequently criticised by academics, trade unions and civil society groups (Megersa, 2019) because of the power imbalance in the buyer/supplier relationship imposed by retail brands. Typically, large international fast fashion brands operating in the global mass market dictate the terms and costs of supply, leaving little room for negotiation on the part of the suppliers, due to the high level of competition in the manufacturing sector, and considerable consumer demand for low-cost fast fashion often leaving the supplier at the mercy of the brands or disconnected from conversations about efficient textile and garment waste management flows.

Relational capital and sustainable supply chains: Collaborative solutions

In trying to assure the effective implementation of a sustainable supply chain, the competitive advantage for some resides in developing synergetic relations emerging from investment in physical and human assets and complementary resources, commitment, shared knowledge, and assurance of effective governance. All of this can be cemented by closer working relationships (Touboulic and Walker, 2015) based on "trust, commitment, complementary resources and capabilities, and investment in supplier development activities" (Brun et al., 2020, p. 4).

In a fragmented supply chain, collaboration appears to be the key to ensuring more sustainable practices. Ideally, collaboration can be either vertical, involving suppliers and customers, or horizontal, with competitors and third parties such as NGOs and trade organisations, working together to fulfil shared goals and reciprocal benefits. Given that the brand owner is regarded as having extended responsibility for the provenance of their products, typically this has been founded on unequal partnerships whereby the brand driving the order has assessed and monitored the situation using audits, for example, and if problems in the performance and delivery of sustainable output were identified, the decision has been either to locate alternative suppliers or invest in enhanced performance (Brun et al., 2020). This relationship oscillates between power and trust and from the restrictions of compliance to more enlightened collaboration. Ideally, it is based on the investment in knowledge capital, information exchange, training and education in addition to generating economic capital to elevate and commit to the assured sustainable operations across the globally dispersed garment and textile supply and value chain based on social and environmental practices.

Country of origin case: Bangladesh textile-apparel industry

The application of a design-driven innovation ecosystem approach to problem solving and policy generation, for both public and private sector organisations, is recognised as a useful driver in stimulating innovation policy (Whicher and Walters, 2017). This approach is based on integrating stakeholder agendas and identifying shared priorities founded on innovative design thinking and strategic operational management. Recent initiatives locating sustainability solutions for fashion and textiles in circular economic models appear to be applying with this approach.

As the world's second largest exporter of ready-made garments after China, Bangladesh accounts for 8% of global textile production (Berg et al., 2021), constituting over 80% of total export earnings equivalent to 33 billion USD with a 6% market share of global apparel exports in 2019 employing over 4 million people, while its linked apparel industry employs around 5 million (Swazan and Das, 2022). Employees work for approximately 4621 export-directed factories with the capacity for 432 spinning divisions, 809 weaving divisions, and 246 dyeing/printing/finishing divisions (BGMEA, 2020). Typically, Bangladesh is associated with linear garment production for international 'fast fashion' brands (Swazan and Das, 2021). The 2019-2022 Covid pandemic and potential global economic downturn further highlight the inequities of this global supply chain dependency, as falling demand and cancelled orders resulted in many workers receiving substantial pay cuts or losing their jobs (Kabir et al., 2021; Islam et al., 2020).

As post-Covid production levels reactivate and the demand for fast fashion in the Global North seemingly remains unabated (Ali et al., 2021), these enterprises generate significant levels particularly of upstream cotton-based material waste or 'jhut' across the contingent manufacturing stages, including spinning, knitting/weaving, dyeing, garment construction, and outputs in the form of faulty or dead stock. Cotton yarn, for example, constitutes the main raw material to knit and weave fabrics in Bangladesh of which 1785 million kg was imported in 2019 with 96,077 tons of polyester staple fibre and 53,289 tons of viscose staple fibre, alongside other raw materials including knitted and woven fabrics (Akter et al., 2022).

Even though the price of materials significantly impacts on the cost of textile and clothing manufacturing, intelligence on the exact quantity and type of textile waste and its impact on profitability or its ultimate economic value when processed in the right way remains seemingly hard to establish. One way of locating and mapping the fate of textile waste in Bangladesh is by applying value stream mapping (VSM) approach using a 7-part classification framework to categorise the reasons for, and the types of resulting waste including overproduction, delays, transport, inappropriate processing, excess inventory, unnecessary movement, and textile/garment defects (Hines and Rich, 1997). A recent study investigating Bangladesh's textile waste streams applied the VSM model (Khalid et al., 2014) to locate the main generator and type of textile waste. The study goes on to quantify the "average material consumption, average material waste, and excess inventory" of 17 selected Bangladesh textile factories using semi-structured questionnaires, observations, and mapping of materials stream across types of industry, raw materials and resulting waste and "interprets the information in economic terms to visualise and quantify value lost in the complete textiles-apparel production chain" (Akter et al., 2022, p. 3). Findings from this study suggested that whether spinning (material conversion rate 89%), textile manufacturing (material conversion rate 81%), wet processing (material conversion rate 91%) or garment production (material conversion rate 80%), the waste was largely generated by over-runs, defective processing, surplus inventory and textile/garment imperfections (Akter et al., 2022). These compelling statistical findings suggest that the amount of textile and garment waste in Bangladesh alone is considerable and provokes the real question as to what happens to it and to what extent it contributes or has the potential to contribute to a circular economy? Bangladesh has great potential to change and be more sustainable in its production practices "to lead the next wave of high-growth economies", but also has significant obstacles,

"whose developmental potential is blocked by political infighting and overstretched infrastructure, with business hindered by red tape,

corruption, energy shortages and the ever-present risk of social unrest. The gap between reality and ambition needs to be bridged for Bangladesh to become high growth economy that benefits the population at large" *(Martin and Impact Economy, 2013, p. 6).*

Bridging that gap may rely on favourable trade and export agreements, the implementation of a Generalised System of Preferences, investment in the physical infrastructure and upskilling human capital (Modi and Zhao, 2020).

The need for transparency in a variety of forms across the supply chain appears critical to assuring and scaling sustainable practices and assuring the benefits of circular business models to achieve those ends. It is important to embed traceability into the strategic management of materials, commodities, component parts and workers to account for the environmental and social settings of the manufacturing plant to assure sustainable and ethical production practices (Fashion Revolution, 2016). The ability to apply traceability mechanisms to track the lifecycle of a garment from sourcing and using raw materials, to their disposal and recycling is also priority, albeit an emerging one. This needs more development with added infrastructural technological investment, for the global textile and clothing industry to manage the value chain efficiently and sustainably. It will become more pressing as policy and legislation is implemented in the country of operation for garment and fashion brands requiring demonstrable compliance.

Traceability challenges and opportunities

Cotton is the main textile used in Bangladesh's RMG production (Nadiruzzaman et al., 2021) with imports from the US and Brazil, for example, supplementing domestic production to keep up with demand for its growing fast fashion exports. To reduce reliance on expensive and volatile cotton imports, recycling post-production textile waste offers a solution given that in 2019 half of the 250,000 tonnes produced in fabric mills by its RMG industry comprised 100% cotton waste, with a potential market value of US$100.000 million (Runnel et al., 2017). The textile is easily recycled when in mono-material form, as the technology and the recycling industry already exists (McKinsey, 2022) and also has environmental benefits given that recycled cotton reduces water consumption, CO_2 emissions and the use of chemicals (Ellen MacArthur, 2013). Typically, differentiated cotton-based textile and garment waste, depending on the season, is produced at different stages of the supply chain in Bangladesh. This cotton textile waste ranges from small strips of cut fabric becoming garment fibre after manual sorting, often used as furniture filling or stuffing, to larger discarded fabric panels from pattern cutting and sewing processes, usually remanufactured into new garments and accessories in small boutiques.

The range of list of "spill types" varies in each garment factory with up 70 subcategories of leftovers based on material composition, size of pieces, rationale for rejection from production and segregation methods, which often means that it is difficult to sort and process efficiently by type and often gets bundled together and sold in the aftermarket reducing its potential for supporting a truly circular economy. The journey of this textile waste beyond the factory floor is also difficult to monitor given that it operates in an informal system at local level, and according to the rules of

> *a vast underground market dealing with 'jhut' and excess inventory apparel, known as 'stock lot'.... This market sells stock lot apparel as it is received or after a bit of refurbishing, and newly sewn apparel from excess fabric or cut-fabric wastes...dealers purchase the cutting waste or stock lot from the apparel manufacturers in a hidden manner (i.e., adopting informal practices) (Akter et al., 2022, p. 8).*

Arguably, this informal, latent business operation in using waste and 'jhut' as a resource with a range of applications and embedded value represents some type of circularity in practice for the benefit of the local textile and garment market and its employees. However, this very large, unofficial market appears to flourish and is often hard to access or regulate, despite the illegality of grey market sales practices transgressing the local licensing laws that aim to protect the local market by forbidding the sale of excess inventory fabric or garments from duty free bonded warehouse sources. Hence, underground sales of waste appear to thrive with approximately 15 tons of 'jhut' fibre sorted by upholsterers alone in the local stores with 5-8 workers, largely female, operating in sub-standard conditions for over 14 hours per day for an average salary of US$60-$80 per month with little care for the health, safety and well-being of employees, which also flouts UN Sustainable Development Goals (UNSDG), notably those relating to ethical and gendered work practices.

Clearly, systematising and making existing large-scale operations more transparent across the entire production stage of the product or garment lifecycle is advisable in the search for a holistic, equitable and financially workable solutions. This can be achieved by more openly managing the significant volumes of waste resulting from various points of textile and garment production in moving towards a circularity solution by adding value and reducing the possibility of landfill dumping or incineration, while stimulating new business possibilities. The solution at ground level in Bangladesh appear to reside in collaborative efforts investing in the use of technology to manage and track data, enabling circular waste recycling systems.

Certainly, Bangladesh is increasingly becoming recognised as one of the largest global centres of upstream recyclable textile waste, particularly cotton, with an estimated value of US$100 million per annum (Repp et al., 2021). One report analysing four Bangladesh based manufacturers commissioned by Reverse Resources (Runnel et al., 2017) estimates that over 500,000 tons of rejected materials are created in Bangladesh per annum, notably from post-industrial textile operations, with a total volume of varied types of waste including yarn, scraps, cut pieces, roll ends, overproduced and substandard pieces and apparel items representing over 47% of raw material input overall.

Grounded leadership in circularity and waste stream management

Some sections of the Bangladesh textile and garment industry, alongside other industries, are exploring new opportunities to reposition themselves as leaders in circularity and waste stream management. In the last decade, emphasis has shifted towards sustainable production issues, particularly for water and energy (Islam et al., 2020). The scale and importance of grounded sustainability developments, and specifically the focus on recycling waste also align with Bangladesh's post-independence status from the 1970s to tackle its vulnerable geographic position regarding the impact of climate change and in managing the consequences of its human displacement issue from its increasingly submerged rural coastlines to the city. Circular economic models are recognised as offering ways of effecting workable sustainable development thinking and practice aligned with the UN SDGs (UN Foundation, 2022). While Bangladesh is a 'Less Developed Country (LDC)', not needing to align with decarbonisation targets, the government appears keen to develop a more sustainable economy with low carbon impact by 2030. There is a policy commitment to use natural resources more sustainably and efficiently to reduce the environmental impact and polluting outputs from manufacturing and supply chains and its dependency on virgin materials in favour of waste management creation in the textile and garment sectors via the reduction, recycling and reuse of materials. This approach also aligns with international legislation and policy planning and underpins the country's socio-economic growth (Shams et al., 2017) and commitment to implementing the UN Sustainable Development Goals and the 10-year plan on Sustainable Consumption and Pattern (SCP), in line with SDG 12 targeting Responsible Production and Consumption in particular.

The transition to circular production and business models in Bangladesh is supported by partnerships and multi-stakeholder initiatives with the Bangladesh government, private sector donors, supporting organisations and development associates. Equally, of ten green factories in the world, six are based in Bangladesh, with over 500 waiting to attain certification from the North American Green Buildings council (Textile Focus, 2022).

With cotton waste being the main Bangladesh waste stream, mechanical cotton recycling, through which cotton is shredded into reusable fibres, has been in use for a long time (Wanassi et al., 2016). One example of mechanical recycling of pre-consumer waste is the large-scale pilot in Bangladesh by the Circular Fashion Partnership (CFP), led by the Global Fashion Agenda (GFA), which adds value by aiming to capture and direct post-production waste back into the production of new textiles, as well as developing solutions for deadstock.

Following a multi-party initiative typical of the innovative design eco-system model (Whicher, 2017), The Circular Fashion Partnership (CFP) headed up by Global Fashion Agenda (GFA) plus Reverse Resources (RR) and the Bangladesh Garment Manufacturers and Exporters Association (BGMEA) alongside selected international fashion brands and textile and garment manufacturers and supported by Partnership for Growth. CFP operates in Bangladesh to enable the textile recycling industry intent on implementing new systems to recycle and re-purpose production fashion waste by securing and re-routing post-production textile and garment waste back into the fashion system where feasible, rather than diverting it into other uses such as upholstery. Additionally, the cross-sectoral project is tasked with sustainably managing the excess inventory of deadstock generated by the Covid-19 crisis, and more generally in working with stakeholders to encourage and facilitate investment and regulation to remove barriers and enhance economic opportunities in Bangladesh's textile and garment sectors. The ultimate intention of this collaborative effort is "to reduce dependency on virgin materials and increase the availability of recycled materials by establishing a long-term, scalable transition to a circular fashion system in garment manufacturing countries" (GFA and McKinsey, 2021). Adapting a design eco-system approach (Niinimäki, 2015), "circular commercial collaborations" are forged with textile and garment producers and brands to produce new products based on repurposed post-industrial waste.

The parties in this coalition create the operational infrastructure, tackle the obstacles and enhance the opportunities to implement viable circular textile and garment production practices. The intended outcome of this initiative is to draw investment to Bangladesh's local recycling operation by refining and showcasing the business opportunity for post-industrial waste recycling so that capacity can be scaled up and new technologies expanded implemented. This is based on the hope that capital sources will offer affordable investment opportunities in clean energy infrastructure.

One of the CFP partners, tracking and trading platform Reverse Resources, calling itself the Uber of textile waste, contributes its proprietary

SaaS platform to systematically track textile waste flows for 360-degree transparency. Based on the belief that the recycling of global textile waste issue is a transparency and access issue rather than a technical one rooted in "a problem of blocked access to waste, incomplete waste data and inflated prices caused by the current waste handling and trading practice" (Runnel et al., 2017). Providing access to transparent and traceable data on textile waste flows enables targeted business decisions based on segmented market reports and analysis. This facilitates a cost-effective, efficient system to develop and communicate demand for products based on recycled materials and heightens access to reliable top quality digitally tracked feedstock for recyclers. The economic argument set out by Reverse Resources suggests that

> *one vertically integrated factory producing 7 million garments per month creates on average ~300 tonnes of leftovers. 500,000 tonnes per year then would refer to 150 factories of that size in Dhaka (2% of all factories)... the total volume of leftovers in Bangladesh is around 400,000 tonnes per year. If these leftovers could be turned into new yarns and garments (considering also more advanced technologies coming available over 5-10 years), this is equivalent to 1.6 billion new garments potentially worth >4 billion USD (from Bangladesh alone) (Runnel et al., 2017).*

Essentially, the RR platform digitally connects-up stakeholders, brands, and recyclers, traversing global market barriers and optimising supply chains by matching supply with demand for recycled feedstock, materials and garments. This match-making service of segregated waste to recycling technology with market demand offers to scale-up the circular economy, while reducing the costs of textile recycling processes. In this way,

> *brands encourage their manufacturing partners to segregate waste and channel it to recycling. Manufacturers receive a fair price for their waste and can transparently report how it has been captured and valorised, a service that can also be presented to other brand customers. Recyclers have access to consistent streams of competitively priced, high-quality feedstock with traceable proof of origin with which they can improve the quality of their recycled material (GFA and McKinsey, 2021).*

Control of the system and quality assurance in addition to continued application of transparency is enhanced by local training in waste management procedures. The pilot scheme also has global applications and aspirations as CFP plans to expand their initiative to Southeast Asian locations, including emerging ventures in Indonesia and Vietnam (Le and Wang, 2017**).** These represent production locations which also represent the competition for Bangladesh's RMG industry in future.

As the global fashion system moves from open loop to closed-loop recycling across the value chain there are still challenges to overcome. In terms of a wish-list for more circular solutions for sustainable textile and garment production from the outset of the product lifecycle at the design planning stage, environmentally friendly raw materials and processing products should be selected to minimise negative environmental impacts and amplify the possibility of an extended product lifetime and its recyclability (Claxton and Kent, 2020; WRAP, 2013).

The mechanical recycling of post-industrial or single blend textile waste using mono fibres such as cotton in both pre- and post-consumer textile waste offers a more straightforward solution compared to other forms of textile waste recycling efforts. Here, textile waste can be retrieved, reused as fibres or offcuts and made into new garments without leaving the same garment manufacturing site, and utilises the existing infrastructure without investing in new technology, while minimising the carbon footprint of production (McKinsey, 2022).

However, beyond mechanical shredding, innovative recycling solutions for multi-textile fibres and garments are emerging on the back of years of R&D and piloting with scale-up capacity (Bürklin and Wynants, 2020). Other initiatives, such as recycled cotton product Circulose (https://circulo.se/en/) or Hong Kong-based yarn spinner Novetex Textiles, working with the Hong Kong Research Institute of Textiles and Apparel (HKRITA), devised The Billie System, based on the mechanical recycling of blended cotton materials (Business of Fashion, 2022). Processing up to 3 tonnes of textile per day, this method is waterless, and no chemical waste is produced. However, there are more problems to overcome with multi-fibre, blended materials where the sorting and collection of worn clothes is still a logistical challenge involving carbon emissions in their transportation to recycling plants, while their physical and material separation can be problematic. Nevertheless, these initiatives are underway, with recycling company Sodra partnering with viscose supplier Lenzing to increase textile waste output by 25,000 tonnes annually by 2025 (BoF, 2021).

Currently, the industry's universal reliance on largely unverified data, that is stored using manual systems without unified standards, is difficult to track and analyse. Industry collaboration, a centralised system of agreed sustainability metrics and supply chain traceability based on enabled by digital technology using blockchain and open-source digital ledgers appear to promise a viable way forward (Sandvik and Stubbs, 2019). Critically, a more transparent, traceable system using universally agreed metrics on water usage for example, tracking software, and an industry-wide, standardised data language based on close cooperation of suppliers and brands across the supply chain cemented by digital data management sharing between the brand, manufacturer and recycler is essential to match supply with demand and enable assurances on sustainable practices to meet environmental and social objectives. Existing initiatives such as SAC's HIGG index and Kering's Taskforce on Climate-Related Financial

Disclosures are useful in this effort but are still too fragmented or limited in application.

One of the key solutions appears to reside in access to transparent data between supply chain stakeholders given that this situation primarily emerges from conflicting business interests compounded by the lack of information sharing and an absence of transparency across supply chains. Circular economic solutions for brands, and suppliers appear to reside in policy-incentivised pre-competitive spaces encouraging digitised open data exchange.

Traceability systems driven by coordinated data input across the textile and garment supply chain are still a work in process. However, there is evidence of emerging platforms enabling the tracking of fibres, textiles, and garments throughout the production process. As most brands only appear to have supplier visibility based on direct contact, there is a critical need for more open, transparent production and sales system, officially and strategically managed, led by traditional data management systems as Reverse Resource's SaaS platform exemplifies. Also, another industry effort is collectively verifying the sustainable source of fibres using blockchain technology. This initiative, TextileGenesis, assigns traceable digital tokens to recycled and organic fibres in partnering with Fashion for Good, Lenzing, Kering, Zalando and the H&M Group (Business of Fashion, 2022).

Furthermore, AI processing to determine demand-led forecasting models within the textile industry 4.0 could focus on reducing excess inventory and rejected stock by enhancing the optimal choice and usage of natural resources and fibres, while scaling back on energy and resource consumption and minimising the carbon footprint of scaled production processes (Ghoreishi et al., 2020). Equally, at the cutting stage, excess trim and materials can be reduced using advanced optimisation and computerised cutting machines (Huang et al., 2021, Li et al., 2021). To exemplify, Reverse Resources cite the success of Bangladesh's large vertically integrated fabric and garment producer Beximco who "by integrating their inventory information from locally maintained excel sheet into the Enterprise Resource Planning (ERP) system...managed to increase the weekly revenue by almost 100k USD" (Runnel et al., 2017).

It should also be a priority to implement effective, efficient, and mindful management of any resulting textile waste emerging at all stages of the production process, from yarn spinning, weaving, and knitting, dyeing and garment creation, aligned with managerial policy and legislation. The need to track and trace the journey of textile waste before, during and after its repurposing to assure an effective circular production model will depend on generating, accessing and sharing vital data and utilising it to match demand with supply between brands and recyclers. This would also enable suppliers and policy makers to refine their systems by measuring and embedding sustainability practices. The former route would track materials handling and management to calculate the value of using sustainable materials or of generating recycled waste and its market worth and scoping new business opportunities in waste management. The latter scenario would entail data generation about production trends, and environmental impact monitoring of carbon emissions for reportage, governance and legislation on the other.

Another grounded solution could take the form of "hidden subsidies" within current linear pricing schemes to encourage suppliers to share their data in digital form, based on more blockchain-based transparent and traceable resource inventories (Runnel et al., 2017). In addition, enhanced material circulation through re-manufacturing by brands of larger sized RMG leftovers in Bangladesh from predictable, technologically tracked waste streams for larger scale upcycled collections is a viable solution (Aus et al., 2021).

Concluding thoughts

Sustained and co-ordinated collaborative efforts and innovative, strategic design-thinking inhabiting a pre-competitive space are required across the design eco-system (Whicher, 2017) involving key parties involved in textile and garment production, including suppliers, merchandisers, policy makers and NGOs are critical to enable concerted efforts to guarantee workable action and to enhance the socio-economic status of manufacturing countries such as Bangladesh, in the short and medium term, as evidenced by the CFP initiative outlined above. Specifically, collaboration within and across industry sectors is vital in terms of establishing a consensus on the need to activate the waste stewardship of materials (McKinsey, 2022). This effort should be based on an understanding of the waste hierarchy, ensuring the minimisation of negative environmental impact via mindless rejection of waste, as opposed to more mindful, planned strategic management, to enhance longevity guided by aligned circular economic guidelines.

The problem is not necessarily about discarded fibres, textiles or garments. It concerns a recognition that they all have use and the ability to operationalise this by the suppliers and brands. So, the implementation of operational system to maximise their value in the waste hierarchy exists. Despite the scale of the challenges facing the textile and garment manufacturing industry in Bangladesh (Akter et al., 2022; Ali et al., 2021) and the wider fashion system, there are reasons to be relatively hopeful as emerging production strategies illustrate how garment producers are recycling fibres and repurposing fabrics and garments to generate renewed value from waste streams, by tracking materials and activating local economies by upscaling recycling capacity and upskilling the workforce, with fashion brands providing market support for and commitment to recycling capacity. In addition, they are tracking data on waste management and exploring options in scaling up nascent recycling, upcycling and reuse of upstream textile waste in different forms, for example, to locate economic benefits from efficient resource management fueling circular economic models aligned with SDG12 Responsible Production and Consumption (Akter et al., 2022).

Emerging and potential collaborative enterprises in Bangladesh also represent circular models aiming to further scale-up the recycling of post-industrial waste that offer capacity-building for future resilience, with social and economic benefits that can be implemented in other manufacturing countries in Asian and Southeast Asian manufacturing bases (Runnel et al., 2017). The vision for efficient material circularity is based on shorter production loops, efficient operational planning, and a more integrated, transparent supply chain.

The workable formula for 'good' textile and fashion production is founded on a pre-competitive (Buchel et al., 2018), collaborative equation of coordinated stakeholder investment of time and resources (economic, technological, information, infrastructure and social) aligned with demonstrable and trackable and required producer responsibilities based on accessible and reliable data generating trust among suppliers and consumers. This relies on the increasing awareness of, and consumer demand for recycled textiles and remanufactured, upcycled garments from the fast fashion consumer and brands closing the loop on supply and demand in producing successful recycled collections (Binet et al., 2019). Yet, ensuring that shared responsible production is aligned with ultimate consumption habits and post-use options are required across the supply chain. Extending and closing the loop by re-evaluating the value and application of textile waste across and beyond the entire global textile and garment supply and value chains driven by strategic partnerships, within innovative, interconnected design eco-systems and pre-competitive spaces, can operate as a unifying differentiator for countries of production and associated stakeholders.

References

Akter, M.M.K., Haq, U.N., Islam, M.M. and Uddin, M.A. (2022). Textile-apparel manufacturing and material waste management in the circular economy: A conceptual model to achieve sustainable development goal (SDG) 12 for Bangladesh. *Cleaner Environmental Systems*, 4: 100070.

Ali, M., Rahman, S.M. and Frederico, G.F. (2021). Capability components of supply chain resilience for readymade garments (RMG) sector in Bangladesh during COVID-19. *Modern Supply Chain Research and Applications*, 3(2): 127-144.

Aus, R., Moora, H., Vihma, M., Unt, R., Kiisa, M. and Kapur, S. (2021). Designing for circular fashion: Integrating upcycling into conventional garment manufacturing processes. *Fashion and Textiles*, 8(1): 1-18.

Berg, A., Chhaparia, H., Hedrich, S. and Magnus, K.H. (2021) *What's Next for Bangladesh's Garment Industry, after a Decade of Growth?* McKinsey & Co., March 25. Available at: https://www.mckinsey.com/industries/retail/our-insights/whats-next-for-bangladeshs-garment-industry-after-a-decade-of-growth (Accessed: 11 August 2022).

Binet, F., Coste-Manière, I., Decombes, C., Grasselli, Y., Ouedermi, D. and Ramchandani, M. (2019). Fast fashion and sustainable consumption. *In:* Muthu, S.S. (Ed.), *Fast Fashion, Fashion Brands and Sustainable Consumption.* pp. 19-35. Singapore: Springer.

Binotto, C. and Payne, A. (2017). The poetics of waste: Contemporary fashion practice in the context of wastefulness. *Fashion Practice*, 9(1): 5-29.

BGMEA. (2021). *Export Performance List.* Available at: https://www.bgmea.com.bd/page/export-performance-list.n Narayanganj (Accessed: 24 February 2022).

Bridge, A. (2022). Greenwashing claims cause fashion brands to pause sustainable Higg Index tool. *PCIAW News.* Available at: https://pciaw.org/higg-index-paused-handm/ (Accessed: 1 August 2022).

British Fashion Council (BFC) (2021). *The Circular Fashion Ecosystem: A Blueprint for the Future.* Available at: https://instituteofpositivefashion.com/uploads/files/1/CFE/Circular_Fashion_Ecosytem_Report.pdf (Accessed: 16 January 2022).

Brooks, A. (2013). Stretching global production networks: The international second-hand clothing trade. *Geoforum*, 44: 10-22.

Brun, A., Karaosman, H. and Barresi, T. (2020). Supply chain collaboration for transparency. *Sustainability*, 12(11): 4429.

Brydges, T. (2021). Closing the loop on take, make, waste: Investigating circular economy practices in the Swedish fashion industry. *Journal of Cleaner Production*, 293: 126245.

Buchel, S., Roorda, C., Schipper, K., Loorbach, D. and Janssen, R. (2018). The transition to good fashion. *DRIFT Report.* Rotterdam: Erasmus University Rotterdam. Available at: https://drift.eur.nl/app/uploads/2018/11/FINAL_report. Pdf (Accessed: 10 June 2022).

Bukhari, M.A., Carrasco-Gallego, R. and Ponce-Cueto, E. (2018). Developing a national programme for textiles and clothing recovery. *Waste Management & Research*, 36(4): 321-331.

Bürklin, N. and Wynants, J. (2020). Opening new opportunities to close the loop: How technology influences the circular economy. *In:* Vignali, G., Reid, L.F., Ryding, D. and Henninger, C.E. (Eds.), *Technology-Driven Sustainability: Innovation in the Fashion Supply Chain.* pp. 219-240. Cham, Switzerland: Palgrave Macmillan.

Business of Fashion (BoF) (2021). *The State of Fashion 2021 Report.* BoF /McKinsey & Company. Available at: https://www.businessoffashion.com/articles/news-analysis/download-the-report-the-state-of-fashion-2021/ (Accessed: 1 February 2022).

Business of Fashion (BoF) (2022). *The State of Fashion Technology 2022 Report.* London: BoF /McKinsey & Company.

Claxton, S. and Kent, A. (2020). The management of sustainable fashion design strategies: An analysis of the designer's role. *Journal of Cleaner Production*, 268: 122112.

Cooper, T. and Claxton, S. (2022). Garment failure causes and solutions: Slowing the cycles for circular fashion. *Journal of Cleaner Production*, 351: 131394.

Corvellec, H., Stowell, A.F. and Johansson, N. (2022). Critiques of the circular economy. *Journal of Industrial Ecology*, 26(2): 421-432.

Coste-Maniere, I., Croizet, K., Sette, E., Fanien, A., Guezguez, H. and Lafforgue, H. (2019). Circular economy: A necessary (r)evolution. *In:* Muthu, S.S. (Ed.), *Circular Economy in Textiles and Apparel*. pp. 123-148. London: Woodhead Publishing.

Department of International Development (2019). *Consumer Pressure, Labour Standards and the Ready Made Garment Sector*. Available at: https://www.gov.uk/research-for-development-outputs/consumer-pressure-labour-standards-and-the-ready-made-garment-sector (Accessed: 14 June 2022).

Domina, T. and Koch, K. (1997). The Textile Waste Lifecycle. *Clothing and Textiles Research Journal*, 15(2): 96-102.

Earley, R. and Goldsworthy, K. (2018). Circular textile design: Old myths and new models. *In:* Charter, M. (Ed.), *Designing for the Circular Economy*. pp. 175-185. London: Routledge.

Ekström, K.M. and Salomonson, N. (2014). Reuse and recycling of clothing and textiles—A network approach. *Journal of Macro-marketing*, 34(3): 383-399.

Ellen MacArthur Foundation (2013). Towards the circular economy. *Journal of Industrial Ecology*, 2(1): 23-44.

Ellen MacArthur Foundation (2017). *A New Textile Economy: Redesigning Fashion's Future*. Available at: https://emf.thirdlight.com/link/2axvc7eob8zx-za4ule/@/preview/1?o) (Accessed: 15 May 2022).

European Commission (2022). *EU Strategy for Sustainable and Circular Textiles*. Brussels: European Commission, Available at: https://eur-lex.europa.eu/legal-content/EN/TXT/?uri=CELEX:52022DC0141 (Accessed: 10 October 2022).

European Environmental Agency (2022). *Textiles and the Environment: The Role of Design in Europe's Circular Economy*. Brussels: EEA, Available at: https://circulareconomy.europa.eu/platform/en/knowledge-type/report (Accessed: 20 June 2022).

Fashion Revolution (2016). *Fashion Transparency Index*. Available at: https://www.lab-of-tomorrow.com/sites/default/files/document/2020-08/FR_FashionTransparencyIndex.pdf (Accessed: 1 August 2021).

Friant, M.C., Vermeulen, W.J. and Salomone, R. (2021). Analysing European Union circular economy policies: Words versus actions. *Sustainable Production and Consumption*, 27: 337-353.

Garcia-Torres, S., Rey-Garcia, M. and Albareda-Vivo, L. (2017). Effective disclosure in the fast-fashion industry: From sustainability reporting to action. *Sustainability*, 9(12): 2256.

Garcia-Torres, S., Rey-Garcia, M., Sáenz, J. and Seuring, S. (2021). Traceability and transparency for sustainable fashion-apparel supply chains. *Journal of Fashion Marketing and Management: An International Journal*, 26(2): 344-364.

Ghoreishi, M., Happonen, A. and Pynnönen, M. (2020). Exploring industry 4.0 technologies to enhance circularity in textile industry: Role of Internet of Things. *Twenty-first International Working Seminar on Production Economics.* pp. 1-16. Innsbruck, Austria, 24-28 February, 2020.

Global Fashion Agenda (GFA) and Boston Consulting (2017). *Pulse of the Fashion Industry 2017-2018.* Available at: https://globalfashionagenda.org/product/pulse-of-the-fashion-industry-2017/ (Accessed: 1 June 2022).

Global Fashion Agenda (GFA) and McKinsey (2021). *Scaling Circularity Report.* Available at: https://globalfashionagenda.org/resource/scaling-circularity-report/

Gutiérrez, W.H. (2018). Acxiom, Red Hat, Actifio, & IBM: A strategic partnership on virtualization and marketing. *Virtu@ lmente,* 6(2): 105-122.

Hines, P. and Rich, N. (1997). The seven value stream mapping tools. *International Journal of Operations & Production Management,* 17(1): 46-64.

Hole, G. and Hole, A.S. (2020). Improving recycling of textiles based on lessons from policies for other recyclable materials: A mini review. *Sustainable Production and Consumption,* 23: 42-51.

Huang, R., Yan, P. and Yang, X. (2021). Knowledge map visualization of technology hotspots and development trends in China's textile manufacturing industry. *IET Collaborative Intelligent Manufacturing,* 3(3): 243-251.

Hvass, K.K. and Pedersen, E.R.G. (2019). Toward circular economy of fashion: Experiences from a brand's product take-back initiative. *Journal of Fashion Marketing and Management: An International Journal,* 23(3): 345-365.

Islam, M.R., Hassan, M.A., Hassan, M.S., Rahman, S. and Nargis, N. (2020). Impact of Covid-19 pandemic on readymade garments (RMG) industry of Bangladesh. *International Journal of Management (IJM),* 11(7): 1125-1132.

Juanga-Labayen, J.P., Labayen, I.V. and Yuan, Q. (2022). A review on textile recycling practices and challenges. *Textiles,* 2(1): 174-188.

Kabir, H., Maple, M. and Usher, K. (2021). The impact of COVID-19 on Bangladeshi readymade garment (RMG) workers. *Journal of Public Health,* 43(1): 47-52.

Kayikci, Y., Kazancoglu, Y., Gozacan-Chase, N., Lafci, C. and Batista, L. (2022). Assessing smart circular supply chain readiness and maturity level of small and medium-sized enterprises. *Journal of Business Research,* 149: 375-392.

Khalid, N.K., Hashim, A.Y. and Salleh, M.R. (2014). On value stream mapping and its industrial significance. *Journal of Industrial and Intelligent Information,* 2(2): 88-90.

Kozlowski, A., Bardecki, M. and Searcy, C. (2012). Environmental impacts in the fashion industry: A life-cycle and stakeholder framework. *Journal of Corporate Citizenship,* 45: 17-36.

Kurpad, M.R. (2014). Made in Bangladesh: Challenges to the ready-made garment industry. *Journal of International Trade Law and Policy,* 80-96.

Le, T.N. and Wang, C.N. (2017). The integrated approach for sustainable performance evaluation in value chain of Vietnam textile and apparel industry. *Sustainability,* 9(3): 477.

Leal Filho, W., Ellams, D., Han, S., Tyler, D., Boiten, V.J., Paço, A. and Balogun, A.L. (2019). A review of the socio-economic advantages of textile recycling. *Journal of Cleaner Production,* 218: 10-20.

Leonas, K.K. (2017). The use of recycled fibers in fashion and home products. *In:* Muthu, S.S. (Ed.), *Textiles and Clothing Sustainability.* pp. 55-77. Singapore: Springer.

Li, X., Wang, L. and Ding, X. (2021). Textile supply chain waste management in China. *Journal of Cleaner Production,* 289: 125147.

Martin, M. and Impact Economy (2013). Creating sustainable apparel value chains. *Impact Economy,* 2(1): 1-41. Available at: https://www.impacteconomy. com/papers/IE_PRIMER_DECEMBER2013_EN.pdf (Accessed: 2 August 2022).

McDonough, W. and Braungart, M. (2013). *The Upcycle – Beyond Sustainability – Designing for Abundance.* New York: Melcher Media.

McDonough, W. and Braungart, M. (2002). *Cradle to Cradle: Remaking the Way We Make Things.* New York: North Point Press.

McKinsey & Co. (2020). *Fashion on Climate: How the Fashion Industry can Urgently Act to Reduce Its Greenhouse Gas Emissions.* Available at: https://www.mckinsey. com/~/media/mckinsey/industries/retail/our%20insights/fashion%20 on%20climate/fashion-on-climate-full-report.pdf (Accessed: 2 May 2022).

McKinsey & Co. (2022). *Scaling Textile Recycling: Turning Waste into Value.* Available at: https://www.mckinsey.com/~/media/mckinsey/industries/ retail/our%20insights/scaling%20textile%20recycling%20in%20europe%20 turning%20waste%20into%20value/scaling%20textile%20recycling%20 in%20europe%20turning%20waste%20into%20value.pdf?shouldIndex=false (Accessed: 2 July 2022).

Megersa, K. (2019). *Consumer Pressure, Labour Standards and the Ready-made Garment Sector.* K4D Helpdesk Report. Brighton, UK: Institute of Development Studies.

Mishra, J.S. and Malhotra, G. (2021). The anatomy of circular economy transition in the fashion industry. *Social Responsibility Journal,* 17(4): 524-542.

Modi, D. and Zhao, L. (2020). Social media analysis of consumer opinion on apparel supply chain transparency. *Journal of Fashion Marketing and Management: An International Journal,* 29(2): 135-149.

Moreira, N. and Niinimäki, K. (2022). *Circular Economy and Fashion: A New Cotton Project White Paper.* Finland: Aalto University Press.

Na, J.H., Evans, M., Zitkus, E. and Walters, A. (2018). Design in action: Understanding the drivers and barriers to strategic use of design for innovation. *In: Academic Design Management Conference: Next Wave.* pp. 127-140.

Nadiruzzaman, M., Rahman, M., Pal, U., Croxton, S., Rashid, M.B., Bahadur, A. and Huq, S. (2021). Impact of climate change on cotton production in Bangladesh. *Sustainability,* 13(2): 574.

Niinimäki, K. (2015). Ethical foundations in sustainable fashion. *Textiles and Clothing Sustainability,* 1(1): 1-11.

Niinimäki, K., Peters, G., Dahlbo, H., Perry, P., Rissanen, T. and Gwilt, A. (2020). The environmental price of fast fashion. *Nature Reviews Earth & Environment,* 1(4): 189-200.

Norris, L. (2019). Waste, dirt and desire: Fashioning narratives of material regeneration. *The Sociological Review,* 6(4): 886-907.

Palm, C., Cornell, S.E. and Häyhä, T. (2021). Making resilient decisions for sustainable circularity of fashion. *Circular Economy and Sustainability,* 1: 651-670.

Payne, A. (2015). Open- and closed-loop recycling of textile and apparel products. *In:* Muthu, S.S. (Ed.), *Handbook of Life Cycle Assessment (LCA) of Textiles and Clothing*. pp. 103-123. Cambridge, UK: Woodhead Publishing.

Reinecke, J., Donaghey, J., Bocken, N. and Lauriano, L. (2019). *Business Model and Labour Standards: Making the Connection*. London: Ethical Trading Initiative. Available at: https://www.ethicaltrade.org/sites/default/files/shared_resources/Business%20models%20%26% 20labour%20standards.pdf (Accessed: 10 August 2022).

Repp, L., Hekkert, M. and Kirchherr, J. (2021). Circular economy-induced global employment shifts in apparel value chains: Job reduction in apparel production activities, job growth in reuse and recycling activities. *Resources, Conservation and Recycling*, 171: 105621.

Reverse Resources (2021). *How Much Does Garment Industry Actually Waste?* Available at: https://reverseresources.net/blog/how-much-does-garment-industry-actually-waste (Accessed: 3 June 2022).

Ross, R. (2004). *Slaves to Fashion: Poverty and Abuse in the New Sweatshops*. Ann Arbor, University of Michigan Press.

Runnel, A., Raihan, K., Castel, N., Oja, D. and Bhuiya, H. (2017). *Creating Digitally Enhanced Circular Economy within Global Supply Chains* [White Paper]. Reverse Resources. Available at: http://www.reverseresources.net/about/white-paper (Accessed: 3 June 2022).

Sandvik, I.M. and Stubbs, W. (2019). Circular fashion supply chain through textile-to-textile recycling. *Journal of Fashion Marketing Management*, 23: 366-381.

Shams, S., Sahu, J.N., Rahman, S.S. and Ahsan, A. (2017). Sustainable waste management policy in Bangladesh for reduction of greenhouse gases. *Sustainable Cities and Society*, 33: 18-26.

Swazan, I.S. and Das, D. (2021). What is Bangladesh's competitive advantage? An exploratory content analysis of the unique firm resources claimed by Bangladeshi apparel export firms. *International Journal of Fashion Design, Technology and Education*, 14(1): 69-77.

Swazan, I.S. and Das, D. (2022). Bangladesh's emergence as a ready-made garment export leader: An examination of the competitive advantages of the garment industry. *International Journal of Global Business and Competitiveness*, 17(1): 1-13.

Textiles Exchange (2021). *Textile Exchange Guide to Recycled Inputs*. Available at: https://textileexchange.org/wp-content/uploads/2021/09/GRS-202-V1.0-Textile-Exchange-Guide-to-Recycled-Inputs.pdf (Accessed: 14 June 2022).

Textiles Exchange (2020). *Preferred Market & Materials Report 2020*. Available at: https://textileexchange.org/app/uploads/2021/04/Textile-Exchange_Preferred-Fiber-Material-Market-Report_2020.pdf. (Accessed: 10 June 2022).

Textile Focus (2022). *Bangladesh now has 165 LEED Certified Green Garment Factories*. Available at: https://textilefocus.com/bangladesh-now-has-165-leed-certified-green-garment-factories/ (Accessed: 10 October 2022).

Touboulic, A. and Walker, H. (2015). Love me, love me not: A nuanced view on collaboration in sustainable supply chains. *Journal of Purchasing Supply Management*, 21: 178-191.

Townsend, K. and Mills, F. (2013). Mastering zero: How the pursuit of less waste leads to more creative pattern cutting. *International Journal of Fashion Design, Technology and Education*, 6(2): 104-111.

United Nations (2015). *The Paris Agreement.* Available at: https://unfccc.int/process-and-meetings/the-paris-agreement/the-paris-agreement (Accessed: 12 May 2022).

United Nations Foundation (2022). *Sustainable Development Goals.* Available at: https://unfoundation.org/what-we-do/issues/sustainable-development-goals/?gclid=EAIaIQobChMIop2Phrqm-wIVB-7tCh14TAo1EAAYAiAA EgJxIPD_BwE (Accessed: 10 November 2022).

Vecchi, A. (2020). The circular fashion framework – The implementation of the circular economy by the fashion industry. *Current Trends in Fashion Technology & Textile Engineering,* 6(2): 31-35.

Wanassi, B., Azzouz, B. and Hassen, M.B. (2016). Value-added waste cotton yarn: Optimization of recycling process and spinning of reclaimed fibers. *Industrial Crops and Products,* 87: 27-32.

Wang, Y. (2010). Fiber and textile waste utilization. *Waste and Biomass Valorization,* 1(1): 135-143.

Whicher, A. (2017). Design ecosystems and innovation policy in Europe. *Strategic Design Research Journal,* 10(2): 117-125.

Whicher, A. and Walters, A. (2017). Mapping design for innovation policy in Wales and Scotland. *The Design Journal,* 20(1): 109-129.

WRAP (2013). *Clothing Longevity and Measuring Active Use.* Banbury, WRAP. Available at: http://www.wrap.org.uk/content/clothing-longevity-measuring-active-use (Accessed: 13 August 2020).

WRAP (2022). *Textiles 2030 Action Plan.* Available at: https://wrap.org.uk/taking-action/textiles/initiatives/textiles-2030 (Accessed: 6 June 2022).

Sorting and Composition Analyses of Discarded Textiles

Kirsti Cura[1]* and Mikko Mäkelä[2]

[1] Aalto University, Finland
[2] VTT Technical Research Centre of Finland Ltd.
e-mail: *kirsti.cura@aalto.fi

1. Introduction

Global fibre production has risen from 51 to 108 million tonnes in the past 20 years (Statista, 2021a). In 2020, natural fibres accounted for 25% of the volume and synthetics for the rest. Typical synthetic fibres in textiles are polyesters and polyamides. Viscose is an example of manmade cellulosic fibres (MMCF). The production volumes of synthetic fibres have grown from 31 to 81 million tonnes in the past 20 years. This number also includes more sustainable MMCFs. In 2020, the market share of synthetic fibres was 62%. The share of polyester alone was 52%. Global fibre production is expected to reach 146 million tonnes by 2030 (Textile Exchange, 2022) and 300 million tonnes by 2050 (Ellen MacArthur Foundation, 2017). There is already a disparity in cotton production compared to global needs. Even though cotton cultivation can be increased to some extent, it is limited mainly due to seasonal weather conditions and land area available for cultivation. Cotton production also uses great amounts of water, pesticides, and other harmful chemicals. This disparity gives MMCFs new opportunities (Felgueiras et al., 2021).

A recent estimate states that by 2030 the gross amount of textile waste will be 8.5–9 million tonnes in the European Union (EU) and Switzerland (McKinsey & Company, 2022). The EU-revised Waste Framework Directive requires EU member states to arrange a separate collection of waste textiles by 2025 (European Union, 2018). In 2019, 2.8 million tonnes of textile waste were separately collected in the EU, most of it being post-consumer textile waste. This amount is expected to reach 4.2–4.5 million tonnes by 2025 with an upcoming separate collection requirement (EURATEX, 2020). This amount will most likely be non-reusables or other items which have low value for the second-hand market. The increased amounts of textile waste call for the need for automated sorting and identification based on fibre type and composition, colour, and structure. They are essential for good quality sorting and a prerequisite for economically viable recycling processes. Near-infrared (NIR) spectroscopy is typically used for textile material identification, but new technologies, such as Raman spectroscopy and hyperspectral imaging, are under development. Designers can play a key role in making the textile and fashion industry more sustainable using circular design strategies. Improvements in product design is the first step to addressing various technical challenges, such as textile identification of mixed fibres, according to the new EU Strategy for Sustainable and Circular Textiles (European Commission, 2022).

As mentioned before, 2.8 million tonnes of textile waste was generated in the EU in 2019 (EURATEX, 2020), 17 million tonnes in the US in 2018 (the United States Environmental Protection Agency, 2021), and 92 million tonnes globally on an average every year (Beall, 2021). The global generation of textile waste is expected to reach 134 million tonnes by 2030 and 150 million tonnes by 2050 (Köhler et al., 2021; Ellen MacArthur Foundation, 2017). In 2020, less than 0.5% of the global fibre market was from pre- and post-consumer textiles (Textile Exchange, 2022). Reliable information on the quantity and quality of textile waste is essential for planning the reuse and recycling of this generated waste. Information on compositions of post-consumer textile waste also varies between regions and seasons. Based on recent estimates, approximately one-fourth of household textile waste consisted of cotton and two-thirds of synthetics (Köhler, 2021; Beasley and Georgeson, 2014). Cotton is currently the most wanted textile fibre for recyclers (Kinden, 2022).

Recycled polyester is the most used recycled fibre (Textile Exchange, 2021). Approximately 15% of the overall polyester production consists of recycled polyester, the majority of which is produced from recycled PET bottles (Statista, 2021b). Recycling bottles to textiles, however, breaks the closed loop of food contact materials, leading to premature generation of PET waste (European Commission, 2022).

2. Product design and fibre composition govern textile recycling

Identification of textile fibres and fibre compositions is closely linked to textile sorting and recycling. Recyclers have different requirements for the fibres depending on their subsequent use. For example, if fibres are used in insulation applications, they can contain different blends and colours. It is estimated that with natural fibres, 5–20% of good quality inputs can be spun into yarn, whereas the amount is much higher with synthetic fibres—but never 100% (Köhler et al., 2021). If a sorted fraction is to be used in chemical recycling, more detailed information is required on textile fibre type and composition (Mäkelä et al., 2021). Some chemical recycling processes can handle some amounts of polyester, elastane, chemicals, or other impurities, whereas others accept 100% monomaterial only. Additional process steps are required to remove impurities which increases costs.

Discarded textiles from consumers are mixed waste streams and contain both reusable textiles and textile waste. The lack of reliable information on textile types and compositions currently limits the ability of recyclers to plan economically viable use for recycled textiles. This generates a "chicken and egg" situation, where the sorters do not know the requirements for sorting and the recyclers cannot be sure if they will receive sufficiently sorted materials for subsequent applications. In addition, the role of product designers is becoming more critical as key decision makers for enabling circular and sustainable products.

2.1 Product design, circular and sustainable products

It has been estimated that 80–90% of the environmental impact of a product is generated and defined in the design stage (Graedel et al., 1995; McAloone and Bey, 2009; Ellen MacArthur Foundation, 2017). Every decision on, for example, raw material selection, manufacturing method, yarn supplier, fabric and garment, manufacturing location, dyeing and finishing chemicals, production and sales volumes, amounts of garments used, and washing affects this environmental impact. Reuse, remanufacturing, recycling, or disposal choices also affect the environmental impact of the product at the end of its lifecycle. Textile supply chains are long, complex, and opaque, and it may be unrealistic to pay attention to every step of the process. The Environment Programme by the United Nation has identified raw material production, various finishing treatments, and consumption as the hotspots which have the largest environmental, economic, and social impacts within the textile supply chain (UNEP, 2020). A more recent report states that material production, including production and finishing of material, and raw material extraction are the main sources of greenhouse gas (GHG) emissions in the apparel sector (Sadowski et al., 2021). Environmental, economic, and social impacts can be considerably decreased by paying most attention to the raw material production, finishing, and consumption steps.

From the design point of view, the lack of material composition data prevents designers from using circular design approaches (Niinimäki and Karell, 2019). Recycling feedstocks can be used as a novel basis to design and develop new materials. This design process should take advantage of the specific properties of the feedstock (Hall, 2021). Dyes can also be recycled either as a chemical compound or as readily dyed fabrics (Määttänen et al., 2019; Hasligner et al., 2019). Fast fashion uses cheaper and lower quality fabrics in larger volumes and manufactures products which are used for a shorter time before being discarded as textile waste (Piippo et al., 2022). The quality of post-consumer waste, especially is decreasing and will likely continue to decrease with the separate collection requirement within the EU (Köhler et al., 2021). A designer can use different sustainable and circular design approaches to increase product lifetime and improve its recyclability. Design for Longevity, Design for Material Recovery, Design for Reuse and Manufacture, Design for Disassembly, and Design for Services are well-known circular design strategies (Niinimäki, 2018). Design for Sorting and Intentional Fashion Design by Recycling Technologies are suggested as new strategies for textile circularity (Karell and Niinimäki, 2019; Niinimäki and Karell, 2019). Textile materials can be reused at the fabric level through remanufacturing or upcycling, whereas recycling takes place at the fibre level. In circular design strategies, a designer is expected to understand the production and recycling processes of clothing and textiles. This knowhow enables designers to exploit Intentional Fashion Design strategies (Niinimäki and Karell, 2019). Circular design approaches can be successfully put into practice by close communication between the designers, sorters, and recyclers.

The relevance of using zippers, buttons, rivets, elastane, blends and different finishing processes as well as chemicals, for example, should be considered from the point of view of recyclability. The Design for Disassembly approach can be used to enable product disassembly at the end of its first lifetime. The main purpose of designing any new products, in any case, should always be based on a genuine need. The requirements of circular design strategies can, however, be in contrast with each other. A typical example is a selection of fabric to be used between a blend of cotton and polyester and 100% cotton. Polycotton blends are widely used as they traditionally perform better compared to cotton by having better durability, making the fabric drapey and easy to iron, but polycotton blends are not always suitable for recycling. Many of the properties brought by polyester are a must for an extended lifetime which gives better sustainability than recycling (Schmidt et al., 2016).

2.2 New possibilities from innovations

Recent studies showed that regenerated man-made cellulose materials can have improved mechanical properties compared to polyester (Moriam et al., 2021; Määttänen et al., 2019). Regenerated cellulose fibres can be several times stronger compared to virgin cotton and could potentially replace synthetic fibres as a reinforcing yarn component (Moriam et al., 2021). This could enable the production of fully cellulose-based materials, which could be chemically recycled as monocellulosics at the end of their lifetime. The greatest potential for reducing the environmental impact of textiles, however, lies in extending their use phase. The greenhouse gas emissions can be decreased by 44% by doubling the use time of garments based on recent estimates (Ellen MacArthur Foundation, 2017).

Mechanical recycling is currently the most common recycling technology for textile waste. Any textile waste can in principle be mechanically recycled by physical manipulation to recover fibres. Fibre length always gets shorter in the process, which means that typically virgin fibres need to be added in spinning to attain good quality yarn. A variety of colours and fibre types often leads to downcycling and premature waste as mixed recycled fibres are not attractive enough for higher value applications. In chemical recycling, textile waste is chemically dissolved to extract different fractions of the components. Depending on the chemical recycling technology, either all or some of the material components can be recovered and reused. Technological development in chemical recycling for both natural and synthetic fibres is booming, and several technologies are expected to be at scale in the near future. A thermo-mechanical recycling process via melting is typically used in recycling some thermoplastic synthetic fibres. There is a significant need to develop recycling alternatives and to find new applications for recycled textiles to avoid incineration and landfilling. Different textile recycling processes are explained in detail in Chapter 8.

3. Sorting options

Different sorting processes can provide the appropriate input for subsequent reuse and recycling processes. Sorting technologies can be classified into four categories: manual sorting, semi-automatic sorting, automated sorting, and information-based sorting (Köhler et al., 2021). The requirements for sorting attributes are determined by the following recycling process and subsequent applications. Recycling textiles to cleaning wipes or mattress stuffing does not require a well-defined input, and in these cases, manual sorting is enough. The raw material for chemical and thermo-mechanical recycling, however, needs to be carefully determined as impurities can affect process control and yield. Chemical recyclers can tune their processes if the incoming material and its variations are well-known. The sorting process can often be the critical step when assessing the economic and environmental costs of the recycling process (Köhler et al., 2021).

Manual sorting based on visual and haptic observations is and will continue to be the main sorting technology globally. It has been estimated that there will be 500x growth in fibre-to-fibre recycling in the next five years (Schweiger, 2022), but not all applications require detailed information on textile fibre types and compositions. However, growing amounts of separately collected waste create even greater pressure to bring more accurate and faster textile identification technologies to support efficient, automated sorting to commercial use. Automated sorting technologies are at an upscaling phase, current capacity being a few thousand tonnes per year in Europe (Köhler et al., 2021). Current identification technologies are known to have limitations which will be discussed later. Many circular design approaches such as Design for Recycling, Design for Remanufacturing, and Design for Reuse require information on material type and composition as this data sets a basis on what kind of products can be designed from said recycled feedstock (Hall, 2021).

3.1 Manual sorting

Manual sorting is widely used by professional recyclers and charity organisations. Materials are identified based on the sorter's experience or the information on the product label. These labels sometimes have inaccurate information, are non-readable, or are entirely missing (WRAP, 2018). The sorting process for consumer textile wastes should start with consumers. Wet, mouldy, or oily textiles need to be removed as they could ruin the whole batch of collected textiles. Textile waste is then pre-sorted to remove dirty, ripped, and torn items from reusable items. The reusable items are further sorted, typically manually. The most labour intensive and hence expensive sorting step is where reusable textile items are manually sorted according to style, colour, season, target group and condition, fabric type and structure, and sometimes even by fabric composition. Large sorting facilities can have up to 350–400 sorting categories (Norup et al., 2018). Reusable items are an important part of the sorters' business model. They typically cover the cost of manual sorting by selling reusable items through their second-hand shops, wholesalers, or other markets. Vintage or second-hand items of higher value can constitute up to 10–15% of the collected textiles (Köhler et al., 2021).

3.2 Semi-automated sorting

Semi-automated sorting uses technological aids for identifying fibre type, composition, colour, or structure. Typical devices are handheld NIR scanners (Beć et al., 2020; Matoha, 2020–2021), some of them connected to mobile phone applications (Senorics, 2021). The challenge with these aids is that the measurements are not fast enough to meet efficiency requirements for sorting. The sorters might therefore need to rely on the information on the product labels even if the labels are known to contain inaccuracies. More information is needed on the effects of inaccurately or falsely labelled textiles in recycling operations. From the economic point of view, it is not beneficial to keep tolerances too high and lose usable input.

3.3 Automated sorting

Automated sorting systems are used to sort non-reusable residuals from manual sorting facilities or collected textile waste that has low reusable content and value (Köhler et al., 2021). Textile items are fed onto a conveyor belt which transports them to an optical instrument and sorts the items based on the measured signals. Non-reusable clean cotton garments and household textiles are excellent and desired inputs for the chemical recycling of cellulosics. The most commonly used identification technology is near-infrared (NIR) spectroscopy which will be explained in more detail in 4.1. From a practical point of view, identification of textile fibre type and composition is carried out from a whole garment/textile item before shredding, and it is a non-destructive measurement. From the point of view of technology, it would be possible to identify T-shirts, vintage garments, and other valuable items by using computer and machine vision technologies. However, these technologies are not yet feasible on an industrial scale. Teaching such a large variety of shapes, patterns, and the concept of vintage to identification software would require a lot of reference attributes and resources.

3.4 Information-based sorting

This sorting method uses radio frequency identification (RFID), near-field communication (NFC), or quick response (QR) tags that are either attached to or printed on the surface of a textile product. The data can be stored either in the tag itself or in a cloud repository. The data can be accessed by reading the tag using a mobile application and can contain information on, for example, the origin of the material, material type and composition, production location, and used chemicals. Retailers use RFID and NFC tags for stock keeping and inventory. Workwear service providers and laundries use RFID tags also for monitoring washing cycles. Circularity.ID by circular.fashion and Circular ID by EON are examples of technology providers that offer solutions for information-based sorting (circular.fashion, 2020; EON Group Holdings Inc., 2021). It is not clear how these physical tags will interfere with recycling. If QR tags are made of the same fabric material as the garment itself and do not contain metal, it should not be a problem. RFID and NFC tags typically contain metal wires and would cause problems in mechanical fibre opening and chemical recycling processes.

4. Identification technologies

The connection between sorting and recycling will remain strong also in the future. Fibre-to-fibre recycling will be increasingly dependent on sufficiently sorted materials to produce high quality outputs. The most common identification method is the naked eye. This accuracy is often enough for low quality applications such as cleaning wipes and insulation, but more sophisticated recycling options require more detailed information about the input. The circular textile business operates at very small profit margins. The best approach should be to produce sufficiently sorted materials for a specific recycling process while taking economic, environmental, and social impacts into consideration.

Discarded textiles vary in composition and quality. The composition of post-industrial textile waste is generally well-known as manufacturers have information on their used materials and volumes. Post-consumer waste is more challenging and various identification methods and technologies are required to ensure appropriate materials for different applications. Consumer education to distinguish resalable textile items from recyclables would be greatly beneficial to obtaining better quality textile waste streams.

4.1 Current status of the identification of textile fibres

Near-infrared (NIR) spectroscopy, colour cameras, and machine vision are currently used technologies for evaluating textile properties. NIR spectroscopy is the most used technology for identifying textile fibres. It studies the overtones and combinations of anharmonic vibrational modes of different molecules (Pasquini, 2018). These modes can be detected in the 780–2500 nm wavelength range and provide information on the fundamental vibrational modes from the mid-infrared range. NIR measurements of textiles are generally performed in diffuse reflectance, and the aim is to estimate the share of incident light absorbed by the fabric across different wavelengths. This generates a spectral fingerprint which is unique for every textile type. The spectral fingerprint measured from a textile sample can then be mathematically compared with a spectral library or a training set of known textile samples to identify which class it belongs to. Evaluating the colour of fabric requires extending the wavelength range to 380–700 nm, which is the visible part of the electromagnetic spectrum.

The chemical information provided by NIR enables distinguishing the most common synthetic and natural fibres, such as polyester, nylon, cotton, viscose, wool, and their blends (Blanco et al., 1994; Cura et al., 2021; Li et al., 2021). Current commercial fibre sorting technologies, Fibersort (Fibersort, 2022) and Siptex (Vinnova, 2022), are based on NIR. Fibersort also uses colour cameras for sorting. NIR spectroscopy has its shortcomings. Identification is a surface measurement, and light has a limited penetration depth in this wavelength range. This can be problematic with coated or multi-layered fabrics (Mäkelä et al., 2020). Thin, loosely knitted fabrics, lace, and dark colours have also been reported to be difficult to identify by measuring an average spectrum from an arbitrary point within the fabric (Cura et al., 2021; Rintala, 2019). Currently used sorting-related identification attributes are summarised in Table 1.

4.2 Future identification technologies of textiles

The identification of textile fibres will continue to see benefits from the development of optical sensors. We anticipate that NIR spectroscopy will remain the most important technology supported by the ease in sample preparation and ongoing efforts in device miniaturisation and the development of imaging sensors with lower costs. Handheld NIR spectrometers are already commercially available (Beć et al., 2020), and promising studies have recently been reported on their use for textile identification (Yan and Siesler, 2018; Rashed et al., 2021). Smaller and cheaper sensors will make these methods accessible to a wider audience within the textile sector.

Modern spectral imaging and machine vision methods provide both spatial and spectral information (Li et al., 2019; Blanch-Perez-del-Notario et al., 2019; Mäkelä et al., 2020; Mäkelä et al., 2021) and reduce the need to control the location of an object on measurement in off-line or on a conveyor belt in on-line conditions. Machine vision generally combines an imaging instrument and a software component with the mathematical algorithms required for identifying different objects or fibre types. Spectral imaging tools complement traditional and portable NIR spectrometers and offer clear benefits for high throughput applications. Industrial sorting systems based on imaging sensors are already commercially available for construction and demolition waste and plastics (Kamppuri et al., 2018; Sormunen and Järvinen, 2021). The global spectral imaging market has been estimated to grow by an annual 15% during 2019–2023 (BCC Research, 2019) and will soon include more and cheaper products to detect the visible and NIR ranges with a single sensor. The principles of machine vision for fibre identification are further summarised in Figure 1.

An important question for the development of optical methods is the level of detail which will be required for future textile identification, sorting, and recycling. For example, will it be sufficient to reliably discriminate the main synthetic and natural fibres? Will some textile materials dominate the

Table 1. Identification attributes in current sorting processes

Sorting attribute	Identification method used at sorters and recyclers	Purpose	Remarks
Fibre type	Sorter's experience Product label NIR	For mechanical, chemical, and thermal recycling	Monomaterials (cotton, polyester, wool, silk, etc.) are easily identified by NIR; regenerated cellulosics can be separated from virgin cellulosics by NIR (Cura et al., 2021), and blends with three or more fibre types are not possible (Köhler et al., 2021)
Fibre composition	Sorter's experience Product label NIR	For mechanical, chemical, and thermal recycling	Blends, especially with more than two different fibres and multilayer structures are impossible (Köhler et al., 2021)
Colour	Visually UV-VIS RBG camera	Dyeing of recyclate can be avoided if colours are sorted well	Dyeing causes a high environmental impact and colour recycling is hence preferred (Haslinger et al., 2019; Zhou et al., 2021)
Structure	Visually Machine vision	Knits are easier to mechanically open compared to woven (Norup et al., 2018; Hall, 2021)	More information is needed on how structure affects textile waste handling

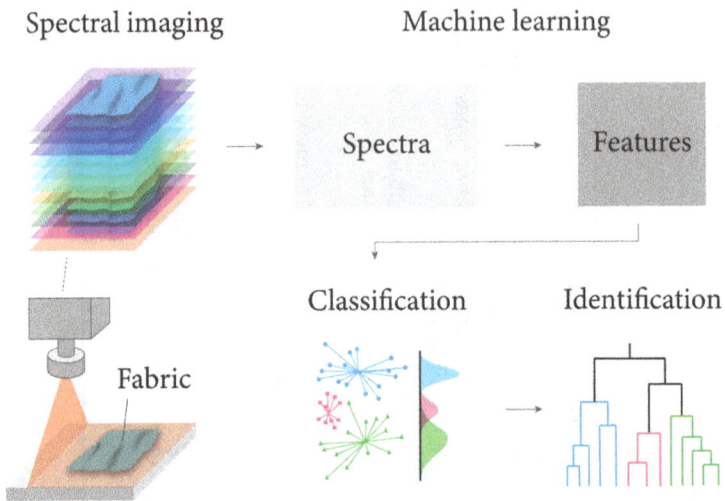

Figure 1. Machine vision can combine imaging spectroscopy and machine learning for identifying textile fibres. A set of spectral images are first measured to determine the spectra of known textile samples and the underlying features in these spectra are used for training a classification model. This model can then be used to determine the fibre types of unknown textile samples. Image credit Mikko Mäkelä and Ella Mahlamäki (VTT Technical Research Centre of Finland Ltd.)

need for developing new identification technology? And will we require more detailed information on the chemistry of different synthetic fibres or the polymer properties of cellulose fibres for conveniently upcycling them to regenerated fibres using chemical methods? This will increase the required complexity and suggests the need to narrow down the variation in textile properties in a stepwise manner. Molar absorptivity also increases as a function of wavelength, making longer wavelengths important for improving the chemical selectivity of optical methods. This can play a significant role in evaluating the properties of synthetic fibres. Synthetic fibres cover over 60% of annual fibre production for clothing applications (Ellen MacArthur Foundation, 2017) and more than one-third of post-consumer waste consists of fibre blends such as polycotton (Ward et al., 2013). The recycling of synthetic fibres needs to be properly addressed to enhance the impact of textile recycling for clothing and other applications. This is needed to fulfil the requirements of increasing recycling rates and the use of recycled textile fibres in the future.

Harmful and hazardous chemicals also need to be identified to phase them out of circulation. These substances cannot currently be reliably identified even in laboratory conditions, not to mention at automated sorting lines. This is a challenging task as there are thousands of chemicals used in the textile industry (Posner and Jönsson, 2014). A new approach was recently proposed to evaluate material composition and contamination with a multispectral approach with artificial intelligence (Rudisch et al., 2021). Rudisch et al. are also working on trend-based sorting and sorting for second-hand use (circular.fashion, 2021).

Information-based sorting technologies including material tracking and tracing will bring an interesting challenge to the identification field. Novel tracking and tracing methods focus on embedding material and product data into textiles in a yarn or fabric production phase, so that the data can be read at the end of the lifecycle. If embedding technologies will be successful, they can change the needs and requirements for different optical identification technologies. Both embedding and data reading technologies are currently in the development phase. We expect that there will be smaller manufacturers, recyclers, and NGOs who will not be able to use these sophisticated technologies. It remains to be seen which technologies will be adopted in the future, and we assume an array of different approaches will be needed.

5. Conclusions, impact on recycling, and lifetime management

Several different fibre-to-fibre recycling technologies are emerging worldwide at various scales. Textile material identification and sorting play a key role in improving the economic viability of these recycling alternatives. More information is required on the feedstock and the effect of its quality in different recycling processes. Recyclers currently have access to good quality feedstocks and are therefore reluctant to accept lower quality textile waste as it can interfere with their production. Almost a third of all textile waste is not currently suitable for fibre-to-fibre recycling, as multi-layered textiles and fibre blends are the main issues (Köhler et al., 2021). We need to develop new applications for discarded textiles, which are not accepted for recycling and currently end up in incineration or landfills. Circular decisions made at the product design phase could considerably decrease the amount of generated textile waste.

More attention should also be directed to extending the lifetime of textile products. Designers have the potential and a great opportunity to improve the lifetime and recyclability of the products using circular design approaches. Material and production choices made by the designer determine the environmental, economic, and social impacts of the product. We need more open discussions between designers, sorters, and recyclers to share information on the challenges, bottlenecks, and opportunities between different actors. Designers also require more information on current and emerging identification and sorting technologies for the development of circular design strategies. Company strategies must support the change towards more circular and sustainable approaches. We need system level thinking across the entire textile value chain.

Acknowledgements

This research was supported by the Strategic Research Council at the Research Council of Finland, Grant no 352616 and no 352698 FINIX consortium.

References

BCC Research (2019). *Hyperspectral Imaging: Technologies and Global Markets to 2023.* Compiled by Gaurav S.G., Report Code IAS135A.
Beall, A. (2020). *Why Clothes are so Hard to Recycle.* BBC (online). Available at: https://www.bbc.com/future/article/20200710-why-clothes-are-so-hard-to-recycle (Accessed: 17 May 2022).
Beasley, J. and Georgeson, R. (2014). *Advancing Resource Efficiency in Europe: Indicators and Waste Policy Scenarios to Deliver a Resource Efficient and Sustainable Europe.* Brussels: European Environmental Bureau.
Beć, K.B., Grabska, J. and Huvk, C.W. (2021). Principles and applications of miniaturized near-infrared (NIR) spectrometers. *Chemistry: A European Journal,* 27(5): 1514-1532. doi: 10.1002/chem.202002838
Blanch-Perez-del-Notario, C., Says, W. and Lambrechts, A. (2019). Hyperspectral imaging for textile sorting in the visible-near infrared range. *Journal of Spectral Imaging,* 8: a17. doi: 10.1255/jsi.2019.a17
Blanco, M., Coello, J., Iturriaga, H., Maspoch, S. and Berthan, E. (1994). Analysis of cotton-polyester yarns by near-infrared reflectance spectroscopy. *Analyst,* 119: 1179-1785. doi: 10.1039/AN9941901779
Circular.fashion (2020). *Enabling the Transformation to Data-driven Circularity in Fashion* (online). Available at: https://circularity.id/ (Accessed: 17 May 2022).
Circular.fashion (2021). *CRTX – Revolutionising Circular Textile Sorting.* AI x Optics x Circularity (online). Available at: https://crtx.ai/ (Accessed: 17 May 2022).
Cura, K., Rintala, N., Kamppuri, T., Saarimäki, E. and Heikkilä, P. (2021). Textile recognition and sorting for recycling at an automated line using near infrared spectroscopy. *Recycling,* 6(11). doi: 10.3390/recycling6010011
Duhoux, T., Maes, E., Hirschnitz-Garbers, M., Peeters, K., Asscherickx, L., Christis, M., Stubbe, B., Colignon, P., Hinzmann, M. and Sachdeva, A. (2021). *Study on the Technical, Regulatory, Economic and Environmental Effects of Textile Fibres Recycling.* Final Report (online). Available at: https://www.ecologic.eu/sites/default/files/publication/2022/50030-study-textile-recycling-web.pdf (Accessed: 17 May 2022).
Ellen MacArthur Foundation (2017). *A New Textiles Economy: Redesigning Fashion's Future* (online). Available at: http://www.ellenmacarthurfoundation.org/publications (Accessed: 17 May 2022).

EON Group Holdings, Inc. (2021). *The Global Data Protocol for Digital Identification* (online). Available at: https://www.eongroup.co/circular-product-data-protocol (Accessed: 01 June 2022).

EURATEX (2020). *ReHubs A Joint Initiative for Industrial Upcycling of Textile Waste Streams & Circular Materials* (online). Available at: https://euratex.eu/wp-content/uploads/Recycling-Hubs-FIN-LQ.pdf (Accessed: 05 August 2022).

European Commission (2022). COMMUNICATION FROM THE COMMISSION TO THE EUROPEAN PARLIAMENT, THE COUNCIL, THE EUROPEAN ECONOMIC AND SOCIAL COMMITTEE AND THE COMMITTEE OF THE REGIONS. *EU Strategy for Sustainable and Circular Textiles.* COM(2022) 141 final (online). Available at: https://ec.europa.eu/environment/publications/textiles-strategy_en (Accessed: 05 April 2022).

European Union (2018). *Directive (EU) 2018/851 of the European Parliament and the Council* (online). Available at: https://eur-lex.europa.eu/legal-content/EN/TXT/PDF/?uri=CELEX:32018L0851&from=EN (Accessed: 09 March 2022).

Felgueiras, C., Azoia, N.G., Gonçalves, C., Gama, M. and Dourado, F. (2021). Trends on the cellulose-based textiles: Raw materials and technologies. *Frontiers in Bioengineering and Biotechnology*, 9(608826). doi: 10.3389/fbioe.2021.608826

Fibersort (2022). *Sorting of Textiles based on Fiber and Color* (online). Available at: https://www.fibersort.com/ (Accessed: 17 May 2022).

Graedel, T.E., Comrie, P.R. and Sekutowski, J.C. (1995). Green product design. *AT&T Technical Journal*, 74(6): 17-25. doi: 10.1002/j.1538-7305.1995.tb00262.x

Hall, C.A. (2021). *Design for Recycling Knitwear: A Framework for Sorting, Blending and Cascading in the Mechanical Textile Recycling Industry.* PhD thesis, University of the Arts London, Chelsea College of Arts, London (online). Available at: https://ualresearchonline.arts.ac.uk/id/eprint/17668/ (Accessed: 05 April 2022).

Haslinger, S., Wang, Y., Rissanen, M., Lossa, M.B., Tanttu, M., Ilen, E., Määttänen, M., Harlin, Al., Hummel, M. and Sixta, H. (2019). Recycling of vat and reactive dyed textile waste to new colored man-made cellulosic fibers. *Green Chemistry*, 21: 5598. doi: 10.1039/C9GC02776A

Kamppuri, T., Heikkilä, P., Pitkänen, M., Hinkka, V., Viitala, J., Cura, K., Zitting, J., Lahtinen, T., Knuutila, H. and Lehtinen, L. (2019). *Tunnistusteknologiat tekstiilien kierrätyksessä* (in Finnish), VTT Technical Research Centre Finland, Ltd. (online). Available at: https://cris.vtt.fi/en/publications/tunnistusteknologiat-tekstiilien-kierr%C3%A4tyksess%C3%A4 (Accessed: 24 March 2022).

Karell, E. and Niinimäki, K. (2019). Addressing the dialogue between design, sorting and recycling in a circular economy. *The Design Journal*, 22(sup1): 997-1013. doi: 10.1080/14606925.2019.1595413

Kinden, T. (2022). Personal information.

Köhler, A., Watson, D., Trzepecz, S., Löw, C., Liu, R., Danneck, J., Konstantas, A., Donatello, S. and Faraca, G. (2021). *Circular Economy Perspectives in the EU Textile Report,* final report (online). Available at: https://op.europa.eu/en/publication-detail/-/publication/08cfc5e3-ce4d-11eb-ac72-01aa75ed71a1/language-en (Accessed: 17 May 2022).

Li, J., Meng, X., Wang, W. and Xin, B. (2019). A novel hyperspectral imaging and modeling method for the component identification of woven fabrics. *Textile Research Journal*, 89: 3752-3767. doi: 10.1177/0040517518821907

Määttänen, M., Asikainen, S., Kamppuri, T., Ilen, E., Niinimäki, K., Tanttu, M. and Harlin, A. (2019). Colour management in circular economy: Decolourization of cotton waste. *Research Journal of Textile and Apparel*, 23(2): 134-152. doi: 10.1108/RJTA-10-2018-0058

Mäkelä, M., Rissanen M. and Sixta, H. (2020). Machine vision estimates the polyester content in recyclable waste textiles. *Resources, Conservation and Recycling*, 161: 105007. doi: 10.1016/j.resconrec.2020.105007

Mäkelä, M., Rissanen, M. and Sixta, H. (2021). Identification of cellulose textile fibers. *Analyst*, 146: 7503-7509. doi: 10.1039/d1an01794b

Matoha (2020-2021). *FabriTell* (online). Available at: https://matoha.com/fabrics-identification (Accessed: 17 May 2022).

McAloone, T.C. and Bey, N. (2009). *Environmental Improvement through Product Development: A Guide*. Danish Environmental Protection Agency (online). Available at: https://orbit.dtu.dk/en/publications/environmental-improvement-through-product-development-a-guide (Accessed: 04 April 2022).

McKinsey & Company (2022). *Scaling Textile Recycling in Europe – Turning Waste into Value* (online). Available at: https://www.mckinsey.com/industries/retail/our-insights/scaling-textile-recycling-in-europe-turning-waste-into-value (Accessed: 05 August 2022).

Moriam, K., Sawada, D., Nieminen, K., Hummel, M., Ma, Y., Rissanen, M. and Sixta, H. (2021). Towards regenerated cellulose fibers with high toughness. *Cellulose*, 28(15): 9547-9566. doi: 10.1007/s10570-021-04134-9

Niinimäki, K. (Ed.) (2018). *Sustainable Fashion in a Circular Economy*. Aalto ARTS Books (online). Available at: https://acris.aalto.fi/ws/portalfiles/portal/32741443/Sustainable_Fashion_in_a_Circular_Economy.pdf (Accessed: 04 August 2022).

Niinimäki, K. and Karell, E. (2019). Closing the loop: Intentional fashion design defined by recycling technologies. *In:* Vignali, G., Reid, L., Ryding, D. and Henninger, C. (Eds.), *Technology-driven Sustainability: Innovation in the Fashion Supply Chain*. pp. 7-25. Cham, Switzerland: Palgrave Macmillan. doi: 10.1007/978-3-030-15483-7_2

Norup, N., Pihl, K., Damgaard, A. and Scheutz, C. (2018). Development and testing of a sorting and quality assessment method for textile waste. *Waste Management*, 79: 8-21. doi: 10.1016/j.wasman.2018.07.008

Pasquini, C. (2018). Near infrared spectroscopy: A mature analytical technique with new perspectives – A review. *Analytical Chimica Acta*, 1026: 8-36. doi: 10.1016/j.aca.2018.04.004

Piippo, R., Niinimäki, K. and Aakko, M. (2022). Fit for the future: Garment quality and product lifetimes in a CE context. *Sustainability*, 14(2): 726. doi: 10.3390/su14020726

Posner, S. and Jönsson, C. (2014). *Chemicals in Textiles – Risks to Human Health and the Environment*. Report from a Government Assignment. Report 6/14, Swedish Chemical Agency.

Rashed, H.S., Puneet, M., Nordon, A., Palmer, D.S. and Baker, M.J. (2021). A comparative investigation of two handheld near-IR spectrometers for direct forensic examination of fibers in-situ. *Vibrational Spectroscopy*, 133: p. 103205. doi: 10.1016/j.vibspec.2020.103205

Rintala, N. (2019). *NIRS Identification of Black Textiles: Improvements for Waste Textiles Sorting.* BSc thesis. LAB University of Applied Sciences (online). Available at: https://urn.fi/URN:NBN:fi:amk-2019062017408 (Accessed: 17 May 2022).

Rudisch, K., Jungling, S., Carrillo Mendoza, R., Woggon, U.K., Budde, I., Malzacher, M. and Pufahl, K. (2021). *Paving the Road to a Circular Textile Economy with AI.* Gesellschaft fur Informatik e.V. (GI) GI. (Hrsg.): INFORMATIK 2021, Lecture Notes in Informatics (LNI), Gesellschaft fur Informatik, Bonn 2021, 313 (online). Available at: https://dl.gi.de/bitstream/handle/20.500.12116/37688/B1-8.pdf?sequence=1&isAllowed=y (Accessed: 04 April 2022).

Sadowski, M., Perkins, L. and McGarvey, E. (2021). *Roadmap to Net Zero: Delivering Science-Based Targets in the Apparel Sector.* World Resources Institute and Apparel Impact Institute. doi: 10.46830/wriwp.20.00004

Schmidt, A., Watson, D. and Roos, S. (2016). *Gaining Benefits from Discarded Textiles. LCA of Different Treatment Pathways.* Copenhagen: Nordic Council of Ministers (TemaNord) (online). Available at: http://norden.diva-portal.org/smash/get/diva2:957517/FULLTEXT02.pdf (Accessed: 04 April 2022).

Schweiger, P. (2022). *Digital Platform to Enhance the Scale-up of a Circular Textile Economy.* Webinar (online). Available at: https://www.canva.com/design/DAE395_4XoI/5H-sm0Rop0A7drQ8qSnLBQ/watch?utm_campaign=designshare&utm_content=DAE395_4XoI&utm_medium=link&utm_source=shareyourdesignpanel (Accessed: 17 May 2022).

Senorics (2021). *Textile Inspection in the Blink of an Eye* (online). Available at: https://matoha.com/fabrics-identification (Accessed: 17 May 2022).

Sormunen, T. and Järvinen, S. (2021). *Report on the State-of-the-art and Novel Solutions in Sorting of Post-consumer Plastic and Packaging Waste.* VTT Technical Research Centre Finland, Ltd. (online). Available at: https://cris.vtt.fi/en/publications/report-on-the-state-of-the-art-and-novel-solutions-in-sorting-of- (Accessed: 24 March 2022).

Statista (2021a). *Worldwide Production Volume of Chemical and Textile Fibers from 1975 to 2020* (online). Available at: https://www.statista.com/statistics/263154/worldwide-production-volume-of-textile-fibers-since-1975/ (Accessed: 17 May 2022).

Statista (2021b). *Recycled and Conventional Polyester Fiber as a Share of Total Production Worldwide from 2008 to 2020* (online). Available at: https://www.statista.com/statistics/1250998/global-share-recycled-polyester-fiber/ (Accessed: 17 May 2022).

Textile Exchange (2021). *Preferred Fiber & Materials, Market Report 2021* (online). Available at: https://textileexchange.org/wp-content/uploads/2021/08/Textile-Exchange_Preferred-Fiber-and-Materials-Market-Report_2021.pdf (Accessed: 17 May 2022).

The United States Environmental Protection Agency (2021). *Textiles: Material-Specific Data* (online). Available at: https://www.epa.gov/facts-and-figures-about-materials-waste-and-recycling/textiles-material-specific-data (Accessed: 17 May 2022).

UNEP (2020). *Sustainability and Circularity in the Textile Value Chain: Global stocktaking* (online). Available at: https://wedocs.unep.org/20.500.11822/34184 (Accessed: 14 February 2022).

Vinnova (2022). *Swedish Innovation Platform for Textile Sorting (SIPTex)* (online). Available at: https://www.vinnova.se/en/p/swedish-innovation-platform-for-textile-sorting-siptex/ (Accessed: 17 May 2022).

Ward, G.D., Hewitt, A.D. and Russell, S.J. (2013). Fibre composition of donated post-consumer clothing in the UK. *Waste and Resource Management*, 166(1): 29-37. Proceedings of the Institution of Civil Engineers.

WRAP (2018). *Fibre to Fibre Recycling: An Economic & Financial Sustainability Assessment* (online). Available at: https://wrap.org.uk/sites/default/files/2021-04/F2F%20Closed%20Loop%20Recycling%20Report%202018.pdf (Accessed: 17 May 2022).

Yan, H. and Siesler, H.W. (2018). Identification of textiles by handheld near-infrared spectroscopy: Protecting customers against product counterfeiting. *Journal of Near Infrared Spectroscopy*, 26: 311-321. doi: 10.1177/0967033518796669

Zhou, J., Zou, X. and Wong, W.K. (2022). Computer vision-based color sorting for waste recycling. *International Journal of Clothing Science and Technology*, 34(1): 29-40. doi: 10.1108/IJCST-12-2019-0190

Textile Recycling Technologies

Pirjo Heikkilä
VTT Technical Research Centre of Finland Ltd.
e-mail: pirjo.heikkila@vtt.fi

Introduction

The globalised textile production with scattered value chains can be quite non-transparent, untraceable and unsustainable from the environmental and social points of view (Kumar et al., 2017; De Brito et al., 2008) Various environmental impacts are caused at different textile production stages, for example, by the use of chemicals, high consumption of water and energy, generation of solid and gaseous wastes, fuel consumption for transportation and use of non-biodegradable packaging materials (European Parliament, 2022; Lee, 2017; Choudhury, 2014). Social responsibility challenges of the global textile industry are related to poor labour conditions, for example, insufficient wage levels, excessive working hours, insufficient health and safety conditions, as well as forced and child labour (Annapoorani, 2017; Padmini and Venmathi, 2012; Holdcroft, 2015; Donato et al., 2020). In recent years many brands have started managing the environmental and social sustainability of their supply chains (Pederssen et al., 2018).

Furthermore, textiles are consumed in an unsustainable way due to mass production and wasteful fast fashion (European Commission, 2022). Textiles are discarded with less use time compared to earlier decades, and, on average, Europeans use nearly 26 kg of textiles, and discard 11 kg of textiles, annually (European Parliament, 2022). According to the new *EU strategy for sustainable and circular textiles* (European Commission, 2022), the priorities in the future sustainable textile system rely on long-lived textile products, which contain recycled fibres and are recyclable. This longevity will be supported by easily available reuse and repair services (European Commission, 2022). In the linear model, *discarded textiles*, i.e., textile products unwanted by the user for one reason or another, generate a vast waste problem, as most of them (87%) are still either incinerated or landfilled (European Parliament, 2022). In a circular economy, however, discarded textiles are *not waste* per se, but there are many ways to use discarded textiles, leading to environmental and socio-economic benefits (Filho et al., 2019).

Both the European waste hierarchy (Directive (EU) 2008/98/EC) and circular strategies (Potting et al., 2017) are emphasising the prevention of waste and reuse (Figure 1). If discarded textile products are clean, unbroken and still visually attractive, they can be *reused* as products. Options for such include second-hand shops, physical and on-line flea markets, various digital platforms and applications, and charity organisations. Textile products having some damage, but are not fully worn out, should be *repaired*, or they could be *repurposed* or their materials can be utilised in *remanufacturing*. All of these activities should be prioritised over recycling.

However, when textile products are not suitable for reuse and materials cannot be utilised in any other way, their materials should be *recycled*. Currently, the majority of textile recycling is focusing on lower-value applications and, thus, referred to often as downcycling. According to Ellen MacArthur foundation (2017) only less than 1% of discarded textiles are recycled back into raw materials of textiles. For lowering the environmental impact of textiles, it is important that recycled textile materials are used to replace virgin raw materials (Dahlbo et al., 2017).

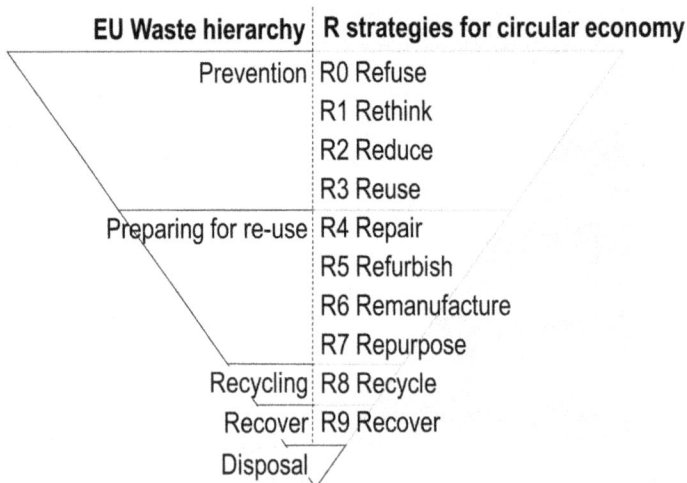

EU Waste hierarchy | **R strategies for circular economy**

EU Waste hierarchy	R strategies for circular economy
Prevention	R0 Refuse
	R1 Rethink
	R2 Reduce
	R3 Reuse
Preparing for re-use	R4 Repair
	R5 Refurbish
	R6 Remanufacture
	R7 Repurpose
Recycling	R8 Recycle
Recover	R9 Recover
Disposal	

Figure 1. Circular economy strategies (Potting et al., 2017) are emphasising two priority categories of EU waste hierarchy (Directive (EU) 2008/98/EC)

In the EU, the separate collection of discarded textiles will be mandatory by 2025 (Directive (EU) 2018/851), and according to the *EU strategy for sustainable and circular textiles,* activities supporting sustainability and circularity of textiles in general will be implemented in the coming years (European Commission, 2020, 2022).

Textile recycling is still very underdeveloped, and successful textile recycling companies are rarer than we have hoped for. There are a number of research activities ongoing for textile recovery, recycling and waste valorisation, and some interesting commercial actors have already emerged. In principle, there are various recycling methods for all main textile fibre, with their own strengths and weaknesses (Kamppuri et al., 2019; Heikkilä et al., 2020). Possible fibre-to-fibre recycling methods are demonstrated for most textile fibre types on a lab scale. However, in practise, recycling of textiles is not a simple task. Textile products are usually not made of a single fibre type, but blends. And, in addition to fibres, textile products also contain other types of materials not suitable for the textile recycling processes. Furthermore, used textile materials can be worn out and contaminated, which makes recycling even more difficult. Sorting and quality assessment are essential stages especially in the processing of mixed textiles from households, which is the most challenging flow to sort and recycle into secondary raw materials for high-value applications. Other textile flows, i.e., side-streams of textile manufacturing processes from the textile industry and unsold items from textile retail, on the other hand, are easier to process, as the material composition is better known and fibre quality is intact (Kamppuri et al., 2019).

This article reviews different recycling possibilities for various kinds of textile materials. Sector 2 focuses on the recovery of discarded textile materials and their pre-processing into secondary raw materials for the textile industry, and Sector 3 on different recycling methods. The main focus is on fibre-to-fibre recycling processes. Sector 4 contains a short summary and a few words about the future.

2. Textile recovery and pre-processing of textile materials for recycling

Discarded textiles need to be collected and sorted so that they can be forwarded into an appropriate utilisation route. Textile products that will be recycled also require more detailed sorting based on the identified fibre type, as well as pre-processing before the actual recycling process is used. The recovery of discarded textiles and pre-processing steps used for recyclable textile fractions are illustrated in Figure 2, and processes are explained shortly in the following paragraphs.

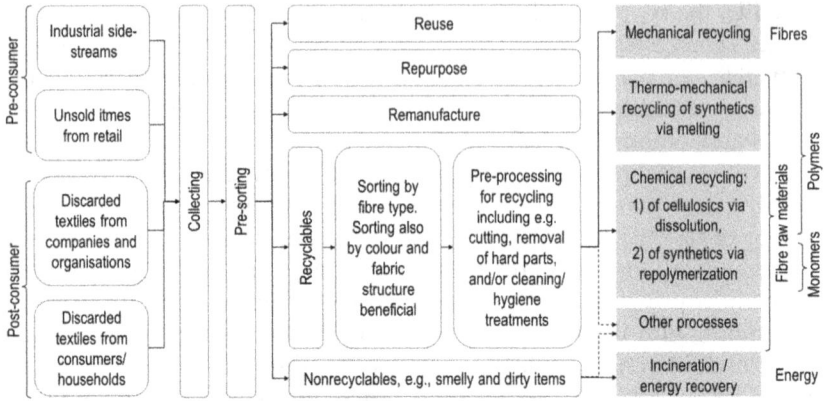

Figure 2. A simplified illustration of the recovery of discarded textiles, pre-processing for textile recycling and recycling routes

There are two main sources for discarded textile flows: (1) *pre-consumer,* i.e., industrial side-streams (sometimes referred to as *post-industrial*) and unsold items from retail, and (2) *post-consumer* (sometimes referred to as *post-consumption*), i.e., discarded textiles from companies/organisations (e.g., hospitals and restaurants) and consumers/households (Fontell and Heikkilä, 2017; Heikkilä et al., 2019; Heikkilä et al., 2020). Textile (waste) flows from these different sources vary in their qualities and quantities. Planning the separate collection of different flows may be challenging. Currently, in the EU members states, there are several different collection systems in operation. In Finland, for example, the collection of household wastes is, for most waste fractions, the responsibility of municipal waste management companies, while the collection and management of waste originating from companies and organisations is the responsibility of commercial waste management companies. This is applied for textile (waste) flows as well, at least for now. France, on the other hand, has adopted the extended producer responsibility (EPR) system for textiles. In the EPR system, producers have a significant responsibility to collect and treat the discarded products. There can be various ways to build up the EPR system, which is why the EU strategy for sustainable and circular textiles proposes harmonised EPR rules for textiles in Europe (European Commission, 2022). When the separate collection of discarded textiles begins in the EU (by the year 2025), there can be slightly different schemes to organise the activities, including combinations of the above-mentioned models.

The sorting of textiles may be done in different stages of the textile recovery process (Heikkilä et al., 2021; LSJH, 2020). In the case of textiles collected from households, *pre-sorting* may be used to remove textile products that can still be reused, repurposed or remanufactured and items that are non-recyclable from the stream. The *material-based sorting* will be done for recyclables, since many recycling processes may be fibre specific. To create new products, fibre composition may be important in order to obtain the desired properties. Furthermore, sorting may be based on colour, since in some cases, colours can be preserved and dyeing is not needed for the next life cycle. Manual sorting is still the dominant procedure for reclaimed textiles, however automated systems and lines are emerging for material-based sorting. Identification and sorting technologies used for textiles are described in more detail in Chapter 8 of this book.

Typically, all textile products that are to be recycled are pre-processed mechanically regardless of the recycling method, and they may also need additional processes. Cutting waste and similar industrial side-streams, for example, do not typically contain accessories, while non-sold items from retailers, for example, are already made into products and may have zippers, buttons, embroidery, tags and other components that may need to be removed prior to recycling. Also, cleaning and/or hygiene treatments may be needed (Heikkilä et al., 2020). A laundry-type process may be done for whole products, but cleaning may be done also in later stages. However, some recycling processes include the use of strong chemicals or high temperatures, which take care of the hygiene issues.

Mechanical pre-processing includes first cutting the textile products into smaller pieces, typically by using guillotine cutting. First, the guillotine cuts textiles into strips after which the direction of the cutter changes 90 degrees, forming a textile shred. In this stage, the pieces containing hard parts, such as buttons, sections of zippers etc., can be removed from shred. Further sorting is also possible at this stage: for example, linings and outer fabrics may have been separated in the shredding process. Textile materials can be recycled at the *fibre level* or at the *fibre-raw-materials level*, and processes include, for example, *mechanical, thermo-mechanical* and *chemical processes*. Mechanical recycling starts from the fibre shred, but for fibre-raw-material recycling, the shred can be cut or ground into smaller pieces (Kamppuri et al., 2019).

An illustration of various possibilities for textile recycling is included in Figure 2. Terminology used in this article is also explained there. It should be noted that the terminology regarding recycling processes is not fixed within the textile sector, and, for example, the thermo-mechanical process is sometimes also referred to as thermal or thermo-plastic process. More confusion may occur as synthetic fibres are plastics. In the textile context, mechanical recycling usually refers to opening the textile structures into fibres and using those fibres in making new products, while in the plastics sector, mechanical recycling refers to the melt processing. Textile processes are explained in more detail in the following sectors.

3. Recycling technologies

Fibre-to-fibre recycling, can be done at the fibre level (mechanical recycling) or the fibre-raw-materials level (thermo-mechanical and chemical recycling) to replace virgin fibres in clothing and textile production. Generally, the existing processes and machinery of the textile industry can be used for all types of recycled fibres. Secondary fibre materials obtained from textile recycling can also be used for the production of nonwovens, composites and other products. Alternative processes are available materials that cannot be easily included into textile-to-textile processes. These include, for example, many laminated and coated textiles, and some blended textiles.

Mechanical recycling is, in principle, suitable for fibre blends, while at the fibre-raw-materials level, recycling with chemical and thermo-mechanical methods are typically polymer specific. In mechanical recycling, the fibre length is reduced and the fibre strength remains unchanged: i.e., worn fibres remain worn, while fibre-raw-materials recycling enables the restoration of properties in fibre spinning, which will follow the recycling process. Fibre-raw-materials recycling at the polymer level enables the restoration of fibre length, and the fibre strength can be restored at least to some extent. However, if the polymer is broken down to monomer the level, the process enables the production of new fibres with properties similar to new ones: i.e., both the polymer and fibre properties are restored. Processes enabling restoration of fibre properties use water and/or chemicals, and therefore are expected to have a higher environmental impact compared to mechanical processing. The challenge is to find a suitable processing method for different types of material, also taking into account the environmental impact of the recycling process, and also finding an optimal high-value application into which material quality allows it to be used.

3.1 Mechanical recycling

Mechanical recycling means textile structures are mechanically torn, opened and unravelled into separate fibres. The possibly unopened material pieces, fibre bundles and debris are removed during the opening process. Recyclate can be blends and mixed material, but the process may be easier to optimise for the processing of a single fibre type and one type of textile structure. The opening process can be continued until the textile is sufficiently opened and suitable for the selected further process. Unnecessary processing should, however, be avoided, as mechanical processing shortens the length of the fibres. In comparison to virgin fibres, mechanically recycled fibres vary greatly in quality as the fibre length is often short (Albrecht et al., 2003). Mechanically recycled fibres can be used in either the spinning of yarns to make fabrics and textile products for the textile industry, or for manufacturing nonwovens and composites (see Figure 3).

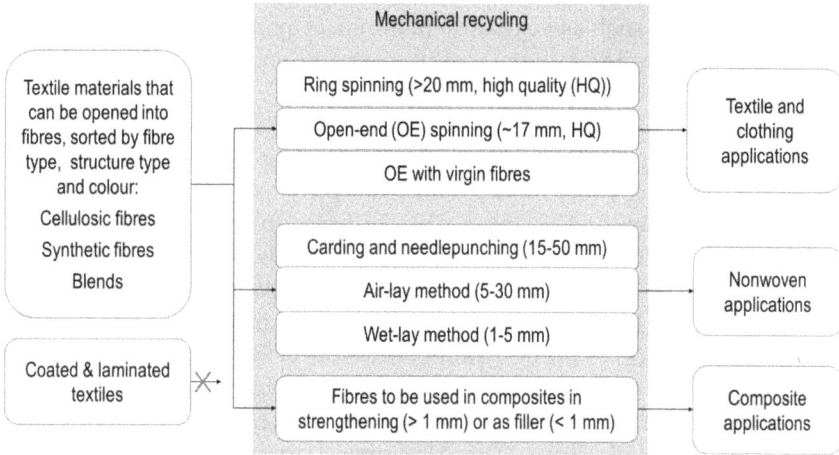

Figure 3. Simplified illustration of the possibilities of mechanical textile fibre recycling

The length of fibre has a key role in strength and durability of the yarns (Aronsson and Persson, 2020), and care must be taken to minimise the loss of fibre length. Furthermore, for good spinnability, the hard parts and textile structure residues also need to be efficiently removed from the opened fibres. The shortest fibres, less than 4-5 mm long, are lost during the processing; fibres that are 12-15 mm long provide bulk and thickness; and fibres longer that 15 mm give spinnability and provide strength and smoothness to the yarns (Klein, 2016). There are various yarn-spinning methods available, and the length of fibre is a key determinant for the method used. Ring spinning can be adjusted based on the fibre length; however, the range of 20-45 mm is considered slightly short for this method. For open-end (OE) spinning, the minimum required fibre length is 17 mm, preferably above 20 mm, and lengths ranging between 25-26 mm are considered good. It is not surprising that OE spinning is favoured for recycled materials, as the fibre length can be shorter compared to ring spinning. In many cases, recycled fibres are blended with longer fibres that have been recycled or with new virgin fibres to ensure high yarn quality (Auranen, 2018; Kamppuri et al., 2019).

Shorter fibres are suitable for nonwovens, which is a group of sheet materials manufactured directly from fibres and bonded into a consolidated structure. The shortest fibres (less than 5 mm long) can be utilised in air-laid nonwovens (i.e., dry papers) and in wet-laid nonwovens.

The air-lay and carding processes are also suitable for longer fibres (length ≥50 mm). Fibres need to be well opened for the wet-laying and carding processes, while in the air-lay process the opening quality is not as critical (Albrecht, 2003). Nonwoven technologies also enable the production of thicker products, such as insulation materials. Opened recycled fibres can be used in composites, either as reinforcement (length >1 mm) or as filler (length <1 mm) (Kamppuri et al., 2019).

Mechanical opening lines are commercially available, and companies offering opening process services and/or providing mechanically recycled fibres for spinning mills include, for example, Rester (FI), Frankenhuis (NL) and Altex Textile Recycling (DE). In order to tackle the problem of the shortening of fibres, new, softer process have also been emerging, such as the Rejuvenation process by PurFi (BE). Other companies, for example Marchi & Fildi (IT), offer yarns at least partly made of mechanically recycled pre-consumer fibres. Another example is Pure Waste Textiles (FI), which has made a *Post waste era* collection with yarns containing 20 per cent of post-consumer cotton fibres blended with other types of recycled fibres (Heikkilä et al., 2019, 2020). Nonwoven production is currently the state-of-the-art for textile recycling and commercialised for multiple types of nonwovens.

3.2 Fibre-raw-materials recycling

Fibre-raw-material recycling can be applied when fibre quality, especially fibre length and strength, need to be restored. It can also be an option in cases where mechanical recycling is not possible, for example, in coated and laminated materials that cannot be opened into fibres. Some processes may also be suitable for fibre blends, even though many of these processes are fibre type-specific due to the specific chemistry used for chemical methods, or due to the used temperature for thermo-mechanical processes.

Fibre-raw-materials recycling can be done at the polymer level or by breaking them down into smaller molecules. Polymer-level recycling can be done via melting (thermo-mechanical processing) or dissolution (chemical), depending on the polymer type (see Figure 4). Thermoformable synthetics can be melted and melt-spun into fibres via thermo-mechanical recycling. Dissolution methods are suitable for cellulose-based fibres and for some synthetic fibres. Synthetic polymers can also be broken down into monomers by chemical and biochemical means, and built back to polymers to be forwarded to fibre spinning processes or manufacturing of other plastic products. These methods are referred to here as chemical recycling methods.

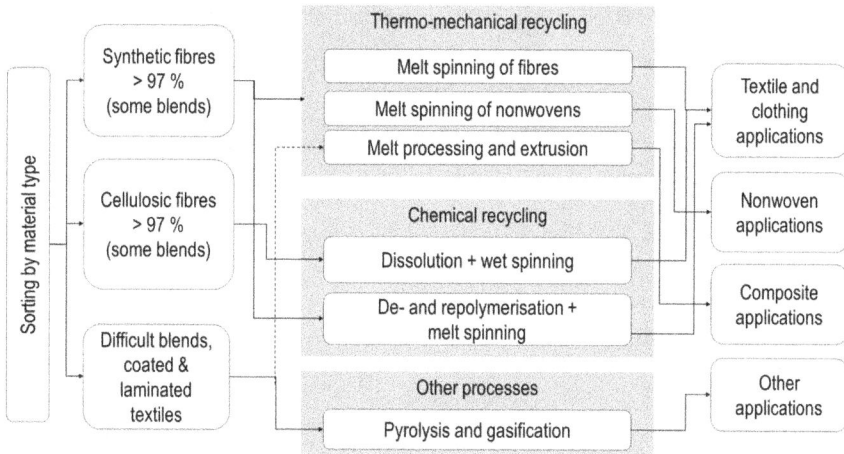

Figure 4. Simplified illustration of possibilities of fibre-raw-materials recycling

3.2.1 *Thermo-mechanical recycling of synthetics – Polymer level*

Thermoplastic polymers are suitable for thermo-mechanical recycling, and they can be melted and melt spun several times. Thermoplastic textile fibres include, for example, polyester, polyamide and polypropylene. Acrylic fibre decomposes close to its melting point and cannot thus be processed this way. Most of the commercially available recycled polyester fibres are currently made with melt-spinning polymer obtained from PET bottles that have been collected from consumers. However, this is not necessarily a preferred method (European Commission, 2022). Processes for recycling polyester fibres from discarded textiles have been studied (e.g., Bascucci et al., 2022) and are slowly emerging.

Challenges for using post-consumer textiles in these processes include physical and chemical changes in polymer occurring during use. More changes, for example in crystallinity and in molecular weight, may occur during thermo-mechanical processing itself. Contaminants may cause chemical reactions that lower the polyester molecular weight, and residues of other materials may weaken the fibres. It is, however, possible to maintain dyes during the process (Saarimäki and Sarsama, 2021; Heikkilä et al., 2021). Molecular weight is a limiting factor for the number of recycling cycles; however, it is possible to valorise polymer melts with additives such as by chain lengtheners (Buccella, 2013; Ozmen et al., 2019) to enable better quality for material, and flame-retarding agents may also affect the reactions during the process (Bascucci et al., 2022). Care must, however, be taken so that additives do not hinder the recyclability of materials in subsequent cycles.

The thermo-mechanical recycling process is excellent for industrial textile side-streams. Companies, such as Nurel (ES), Racidi (IT) and

Fulgar (IT), offer fibres made from remelted PA from production. There are industrial examples for thermo-plastic recycling and upgrading of used synthetic polymers, for example, Cumapol's (NL) processes for polyester. Development work for these processes is ongoing, and fishing nets, for example, have been successfully thermo-mechanically recycled (Mondragon et al., 2020), but those are also suitable for other plastic processes.

Thermoplastic materials can also be used for making composites, where they can be either as fibres or as matrix. Composite can also be made of a cellulosic-synthetic fibre mixture, where the synthetic man-made fibre(s) are melted around cellulose-based staple fibres (e.g., CO or man-made cellulosic fibres (MMCF)) (Kamppuri et al., 2019). Thermo-mechanical methods have shown to be also very promising for the recycling of technical textile materials into high-quality plastic and composite materials (Saarimäki and Sarsama, 2021; Heikkilä et al., 2021).

3.2.2 Chemical recycling of cellulosics – Polymer level

Cellulose-based fibres, such as cotton, flax, viscose and lyocell, can be dissolved and spun into MMCFs. Impurities, such as silicates and metals, can be chemically removed from grinded fibres, and a coloured fraction can be bleached. Cellulosic fibres are dissolved, and the cellulose solution is pressed through h0oles in a nozzle into a spinning bath, where fibres are formed by precipitation. The dissolution and spinning processes vary slightly, but in principle, they are the same methods used for making primary fibres from wood-soluble pulp. The chain length of the cellulose molecules of cotton is higher than that of the dissolving pulp (de Silva, 2017), so the wear of fibres is typically not a problem in cotton recycling.

Typically, raw materials entering the chemical recycling process should have high-cellulose-fibre contents (between 97% and 98%) to reach MMCF; however, methods for higher-mixing-ratio blends are developed actively. Dissolution may be used for the separation of blends by dissolving the cotton and filtering other fibres out from the cellulose solution (De Silva, 2014). Filtration may, however, increase the cost of the process. Some fibres that are dissolved with cotton are also more difficult to remove, for example, elastane. Regarding colour preservation, it has been demonstrated that the colour of fibres can be saved during chemical recycling (Ma et al., 2020). If needed, colour and fibre-finish removal steps may also be included into recycling processes (Wedin et al., 2018).

Recycling of cotton has been demonstrated with the main commercial MMCF processes, i.e., viscose (e.g., Wedin et al., 2018) and lyocell processes (e.g., Haule et al., 2016; Björquist et al., 2018), and emerging processes, including cellulose carbamate (Paunonen et al., 2019), Biocelsol (Vehviläinen et al., 2018, as cited in Vehviläinen et al., 2020),

Ioncell technologies (Asaadi et al., 2016) and mixed solvent process (Ma et al., 2019).

Traditional MMFC companies, like Lenzing (AU) and Kelheim (DE), have introduced products utilising recycled pulp as raw material, but within the last ten years, new companies have also been founded around this business, including, for example, Renewcell (SE), Infinited Fiber Company (FI), Evrnu (US) and SaXcell (NL).

3.2.3 Chemical recycling of synthetics – Monomer level

Various chemical recycling methods have been developed to recycle synthetic fibres, such as polyester, nylon and acrylic. In the process, the polymer chain is broken down into monomers or other molecules, and then re-polymerised into a polymer of preferred length. Such repolymerisation methods are commercially available, for example, for polyamide and polyester. Polyamide 6 can be recycled via ring closing repolymerisation (Alberti et al., 2019); polyamide 6.6 by glycolysis and amino-glycolysis processes (Datta, 2018); and polyester, for example, via glycolysis (e.g., Sert et al., 2019). Other process alternatives include alcoholysis, hydrolysis, aminolysis and ammonolysis (Raheem et al., 2019).

Monomer-level recycling processes typically involve cleaning, colour removal and separation of different molecules obtained in a depolymerisation process. Therefore, purity and quality of the polymers can be restored. Many of the processes are not economically viable yet, however, especially if extensive cleaning steps are used to recycle slightly contaminated textiles. The environmental impacts of such processes may also not be known. Chemical recycling methods open new possibilities for upcycling and making products similar to those made of new polymers. Processes are principally the same as are used for plastics recycling; however, additive chemistry and contaminants of textile waste are somewhat different compared to typical plastics products, such as packaging materials or bottles. Therefore, cleaning and pre-processing might be different (Kamppuri et al., 2019).

Chemical monomer-level recycling has been available for over two decades (Paszun et al., 1997). The first company to use this on an industrial scale was Teijin (JP). New commercial actors have recently joined this business, for example, Aquafil (IT) for the recycling of polyamide (Econyl process) and Carbios (FR) for utilising enzymatic recycling process for polyester.

3.3 Other methods

For materials that cannot be sustainably recycled by the means described in the previous chapters (e.g., various alloy materials and dirty fractions), thermal conversion processes may provide an option. In these processes, the polymer structure is broken down in thermal conversion into short-chain hydrocarbons or other molecular structures. The processes of thermal conversion are pyrolysis and gasification. They are very similar processes to each other but produce different types of finished products. Pyrolysis produces mostly liquid products containing solid carbon of 15-25% and gaseous compounds of 10-20%. Gasification typically produces approximately 85% of the gaseous final product, 10% of solid carbon and 5% of fluid. The characteristics of the raw materials entering into the thermal conversion affect the characteristics of the finished product obtained. Such products may be used by the chemical industry (Kamppuri et al., 2019).

Incineration is also a thermal process, where only thermal energy from the thermal degradation of the material is utilised. The calorific value of the polypropylene, polyethylene and polyester is high (more than 40 MJ/kg) and of the same class as the calorific value of the fuel oil. The incineration of mixed waste does not reach such high readings, as part of the thermal energy is used to dry the moisture contained in mixed waste (Kamppuri et al., 2019).

4. Summary and future prospects

In the future, circular economy textiles must be produced, used and (re)cycled in a sustainable way. While longevity and re-use of the products should be emphasised, we should also be able to deal with the textile waste and to recycle that to be used as raw material for the textile industry. *Post-consumer* materials from organisations and companies using textiles are worn, but these may be known and controlled textile flows; while discarded textiles from household and consumers are mixed, unknown and therefore the most challenging to sort and recycle into secondary raw materials that could be used in high-value applications. Mechanical recycling is applicable when products are unusable, but fibre quality is good and fibre length sufficient for the intended purpose. In order to replace virgin materials in textile products, mechanically recycled fibres must be long enough for the spinning of yarns. Spinning uses longer fibres, while shorter ones can be used for nonwoven and other lower-value applications.

If mechanical recycling is not feasible, and fibre length needs to be restored, there are options for fibre-raw-materials recycling. These include several processes at different levels. It should be noted that the more we process, the more we might cause environmental impacts. Polymer-level recycling includes thermo-mechanical melt-processing of synthetics and chemical recycling (dissolution) of cellulosics and blends. Chemical monomer-level recycling of synthetics and blends also enables restoring polymers. Most of these processes are polymer specific, but some of them are capable of handling blends as well.

Mechanical recycling is still a dominant method, while fibre-raw-materials recycling, such as chemical processes, are emerging but mostly still in pilot scale. There are multiple technological challenges related to recycling technologies themselves. Future developments are expected to focus on such chemical processes , for example, the processing of material blends (Circle Economy, 2020). Development is also needed related to textile collection systems, new identification and sorting technologies and digitalisation.

In addition to technological challenges, non-technological-affecting factors, for example, environmental law and policies also need to be considered (Damayanti et al., 2021; Dissayanake and Weerasinghe, 2021). To make recycling more efficient, we would benefit from the development of a textile classification system. Classification could be based on chemical groups and bonds that form the backbone of the polymers (Harmsen et al., 2021). However, it would be important to also understand the condition of fibre and polymer wear and tear in order to enable more efficient recycling. This is a challenge for the further development of identification technologies. In addition, recognition of contaminants including fibre finishes, would benefit recycling process control and make recycling safer. Furthermore, better understanding is needed on the sustainability of different recycling processes, since from an environmental point of view it would be beneficial to use low-impact recycling processing, such as mechanical process, instead of more water-, chemical- and/or energy-intensive processing, whenever possible.

Acknowledgements

This article was mainly based on work carried out in the Telaketju Tekes project (e.g. Kamppuri et al., 2019) complemented with a literature review carried out in the Telaketju 2 and Telavalue projects. I would like to thank the project groups from all of the Telaketju joint projects and Business Finland for the funding.

References

Alberti, C., Figueira, R., Hofmann, M., Koschke, S. and Enthaler, S. (2019). Chemical recycling of end-of-life polyamide 6 via ring closing depolymerization. *Chemistry Select*, 4: 12638 (online) https://doi.org/10.1002/slct.201903970

Albrecht, W., Fuchs, H. and Kittelmann, W. (Ed.) (2003). *Nonwoven Fabrics. Raw Materials, Manufacture, Applications, Characteristics, Testing Processes.* Weinheim: Wiley-VCH Verlag GmbH & Co. KGaA.

Annapoorani, S.G. (2017). Social sustainability in textile industry. *In:* Muthu, S. (Ed.), *Sustainability in the Textile Industry. Textile Science and Clothing Technology.* Singapore, Springer (online) https://doi.org/10.1007/978-981-10-2639-3_4

Aronsson, J. and Persson, A. (2020). Tearing of post-consumer cotton T-shirts and jeans of varying degree of wear. *Journal of Engineered Fibers and Fabrics,* 1: 1-9 (online) https://doi.org/10.1177/1558925020901322

Asaadi, S., Hummel, M., Hellsten, S., Härkäsalmi, T., Ma, Y., Michud, A. and Sixta, H. (2016). Renewable high-performance fibers from the chemical recycling of cotton waste utilizing an ionic liquid. *ChemSusChem,* 9(22): 3250-3258 (online) http://dx.doi.org/10.1002/cssc.201600680

Auranen, A. (2018). Tekstiilijätteestä mekaanisesti kierrätetty kuitu ja sen soveltuvuus eri prosesseihin. Metropolia University of Applied Sciences. Bachelor Thesis (online) http://www.theseus.fi/bitstream/handle/10024/153658/Auranen_Anneli.pdf?sequence=1&isAllowed=y 15/06/2022

Bascucci, C., Duretek, I., Lehner, S., Holzer, C., Gaan, S., Hufenus, R. and Gooneie, A. (2022). Investigating thermomechanical recycling of poly(ethylene terephthalate) containing phosphorus flame retardants. *Polymer Degradation and Stability,* 195: 109783 (online) https://doi.org/10.1016/j.polymdegradstab.2021.109783

Björquist, S., Aronsson, J., Henriksson, G. and Persson, A. (2018). Textile qualities of regenerated cellulose fibers from cotton waste pulp. *Textile Research Journal,* 88(21): 2485-2492 (online) https://doi.org/10.1177/0040517517723021

Buccella, M., Dorigato, A., Caldara, M., Pasqualini, E. and Fambri, L. (2013). Thermo-mechanical behaviour of polyamide 6 chain extended with 1,1'-carbonyl-bis-caprolactam and 1,3-phenylene-bis-2-oxazoline. *Journal of Polymer Research,* 20: 225 (online) https://doi.org/10.1007/s10965-013-0225-2

Choudhury, A.K.R. (2014). Environmental impacts of the textile industry and its assessment through life cycle assessment, Chapter 1. *In:* Muthu, S.S. (Ed.), *Roadmap to Sustainable Textiles and Clothing.* pp. 1-40. Singapore: Springer.

Circle Economy (2020). *Recycled Post-consumer Textiles: An Industry Perspective* (online) https://www.nweurope.eu/media/9453/wp-lt-32-fibersort-end-markets-report.pdf 13/06/2022

Dahlbo, H., Aalto, K., Eskelinen, H. and Salmenperä, H. (2017). Increasing textile circulation: Consequences and requirements. *Sustainable Production and Consumption,* 9: 44-57 (online) https://doi.org/10.1016/j.spc.2016.06.005.

Damayanti, D. Wulandari, L.A. Bagaskoro, A. Rianjanu, A. and Wu. H.-S. (2021). Possibility routes for textile recycling technology. *Polymers,* 13(21): 3834 (online) https://doi.org/10.3390/polym13213834

Datta, J., Błażek, K., Włoch, M. and Bukowski, R. (2018). A new approach to chemical recycling of polyamide 6.6 and synthesis of polyurethanes with recovered intermediates. *Journal of Polymers and the Environment,* 26: 4415-4429 (on-line) https://doi.org/10.1007/s10924-018-1314-4

De Brito, M.P., Carbone, V. and Blanquart, C.M. (2008). Towards a sustainable fashion retail supply chain in Europe: Organisation and performance.

International Journal of Production Economics, 114(2): 534-553 (online) https://doi.org/10.1016/j.ijpe.2007.06.012.

De Silva, R., Wang, X. and Byrne N. (2014). Recycling textiles: The use of ionic liquids in the separation of cotton polyester blends. *RSC Advances*, 4: 29094-29098 (online) https://pubs.rsc.org/en/content/articlepdf/2014/ra/c4ra04306e 13/06/2022

De Silva, R. and Byrne, N. (2017). Utilization of cotton waste for regenerated cellulose fibres: Influence of degree of polymerization on mechanical properties. *Carbohydrate Polymers*, 174: 89-94 (online) https://doi.org/10.1016/j.carbpol.2017.06.042

Directive 2008/98/EC of The European Parliament and of The Council of 19 November 2008 on waste and repealing certain Directives (online) https://eur-lex.europa.eu/legal-content/EN/TXT/?uri=celex%3A32008L0098

Directive (EU) 2018/851, of the European Parliament and of The Council of 30 May 2018 amending Directive 2008/98/EC on waste (online) https://eur-lex.europa.eu/legal-content/EN/TXT/PDF/?uri=CELEX:32018L0851&from=E 15/06/2022

Dissanayake, D.G.K. and Weerasinghe, D.U. (2021). Fabric waste recycling: A systematic view of methods, applications and challenges. *Materials Circular Economy*, 3: 24 (online) https://doi.org/10.1007/s42824-021-00042-2

Donato, C., Buonomo, A. and De Angelis, M. (2020). Environmental and social sustainability in fashion: A case study analysis of luxury and mass-market brands. *In:* Muthu, S. and Gardetti, M. (Eds.), *Sustainability in the Textile and Apparel Industries. Sustainable Textiles: Production, Processing, Manufacturing & Chemistry.* Cham: Springer (online) https://doi.org/10.1007/978-3-030-38532-3_5

Ellen MacArthur Foundation (2017). *A New Textiles Economy: Redesigning Fashion's Future* (online) https://ellenmacarthurfoundation.org/a-new-textiles-economy 13/06/2022

European Commission (2020). Directorate-General for Communication, *Circular Economy Action Plan: For a Cleaner and more Competitive Europe.* Publications Office, 2020 (online) https://data.europa.eu/doi/10.2779/05068

European Commission (2022). EU Strategy for Sustainable and Circular Textiles. Brussels, 30.3.2022 COM(2022) 141 final (online) https://eur-lex.europa.eu/resource.html?uri=cellar:9d2e47d1-b0f3-11ec-83e1-01aa75ed71a1.0001.02/DOC_1&format=PDF

European Parliament (2022). The impact of textile production and waste on the environment (infographic) Article 20201208STO93327 (online) Available: https://www.europarl.europa.eu/news/en/headlines/society/20201208STO93327/the-impact-of-textile-production-and-waste-on-the-environment-infographic 18/05/2022 22/06/2022

Filho, W.L., Ellams, D., Han, S., Tyler, D., Boiten, V.J., Paço, A., Moora, H. and Balogun, A.-L. (2019). A review of the socio-economic advantages of textile recycling. *Journal of Cleaner Production*, 218: 10-20 (online) https://doi.org/10.1016/j.jclepro.2019.01.210

Fontell, P. and Heikkilä, P. (2017). *Model of Circular Business Ecosystem for Textiles.* VTT Technical Research Centre of Finland. VTT Technology No. 313 (online) https://publications.vtt.fi/pdf/technology/2017/T313.pdf

Harmsen, P., Scheffer, M. and Bos, H. (2021). Textiles for circular fashion: The logic behind recycling options. *Sustainability*, 13: 9714. (online) https://doi.org/10.3390/su13179714

Haule, L.V., Carr, C.M. and Rigout, M. (2016). Preparation and physical properties of regenerated cellulose fibres from cotton waste garments. *Journal of Cleaner Production*, 112(5): 4445-4451 (online) https://doi.org/10.1016/j.jclepro.2015.08.086

Heikkilä, P., Kamppuri, T., Saarimäki, E., Pesola, J., Alhainen, N. and Jetsu, P. (2019). Recycled Cotton Fibres in Technical and Clothing Applications. 4th International Conference on Natural Fibers Porto/Portugal, 1-3 July 2019.

Heikkilä, P., Määttänen, M., Jetsu, P., Kamppuri, T. and Paunonen, S. (2020). *Nonwovens from Mechanically Recycled Fibres for Medical Applications.* VTT Technical Research Centre of Finland. VTT Research Report No. VTT-R-00923-20 (online) https://cris.vtt.fi/en/publications/nonwovens-from-mechanically-recycled-fibres-for-medical-applicati

Heikkilä, P., Cheung, M., Cura, K., Engblom, I., Heikkilä, J., Järnefelt, V., Kamppuri, T., Kulju, M., Mäkiö, I., Nurmi, P., Palmgren, R., Petänen, P., Rintala, N., Ruokamo, A., Saarimäki, E., Vehmas, K. and Virta, M. (2021). *Telaketju – Business from Circularity of Textiles.* VTT Technical Research Centre of Finland. VTT Research Report No. VTT-R-00269-21 (on-line) https://cris.vtt.fi/en/publications/telaketju-business-from-circularity-of-textiles

Holdcroft, J. (2015). Transforming supply chain industrial relations. *International Journal of Labour Research* 7(1-2): 95-104 (online) https://labordoc.ilo.org/discovery/delivery/41ILO_INST:41ILO_V2/1268159920002676 15/06/2022

Kamppuri, T., Pitkänen, M., Heikkilä, P., Saarimäki, E., Cura, K., Zitting, J., Knuutila, H. and Mäkiö, I. (2019). Tekstiilimateriaalien soveltuvuus kierrätykseen, VTT Tutkimusraportti VTT-R-00091-19 (online) https://cris.vtt.fi/en/publications/tekstiilimateriaalien-soveltuvuus-kierr%C3%A4tykseen

Klein, W. (2016). *The Rieter Manual of Spinning.* Volume 1: Technology of Short-staple Spinning. Rieter Machine Works Ltd.

Kumar, V., Agrawal, T.K., Wang, L. and Chen, Y. (2017). Contribution of traceability towards attaining sustainability in the textile sector. *Textiles and Clothing Sustainability*, 3: 5 (online) https://doi.org/10.1186/s40689-017-0027-8

Lee, K.E. (2017). Environmental sustainability in the textile industry. In: Muthu, S. (Ed.), *Sustainability in the Textile Industry. Textile Science and Clothing Technology.* Singapore: Springer (online) https://doi.org/10.1007/978-981-10-2639-3_3

LSJH (2020). *National Collection of End-of-life Textiles in Finland.* Lounais-Suomen Jätehuolto Oy (online) https://telaketju.turkuamk.fi/uploads/2020/08/0c08d295-national-collection-of-end-of-life-textiles-in-finland_lsjh.pdf 13/06/2022

Ma, Y., Zeng, B., Wang, X. and Byrne, N. (2019). Circular textiles: Closed loop fiber to fiber wet spun process for recycling cotton from denim. *ACS Sustainable Chemistry & Engineering*, 7(14): 11937-11943 (online) https://pubs.acs.org/doi/10.1021/acssuschemeng.8b06166

Ma, Y., Rosson, L., Wang, X. and Byrne, N. (2020). Upcycling of waste textiles into regenerated cellulose fibres: Impact of pretreatments. *The Journal of the Textile Institute*, 111(5): 630-638 (online) https://doi.org/10.1080/00405000.2019.1656355

Mondragon, G., Kortaberria, G., Mendiburu, E., González, N., Arbelaiz, A. and Peña-Rodriguez, C. (2020). Thermomechanical recycling of polyamide 6 from fishing nets waste. *Journal of Applied Polymer Science* (online) https://doi.org/10.1002/app.48442

Ozmen, S.C., Ozkoc, G. and Serhatli, E. (2019). Thermal, mechanical and physical properties of chain extended recycled polyamide 6 via reactive extrusion: Effect of chain extender types. *Polymer Degradation and Stability*, 162: 76-84 (online) https://doi.org/10.1016/j.polymdegradstab.2019.01.026

Padmini, D. and Venmathi, A. (2012). Unsafe work environment in garment industries, Tirupur, India. *Journal of Environmental Research and Development*, 7(1A): 569-575 (online) https://www.academia.edu/24274050/UNSAFE_WORK_ENVIRONMENT_IN_GARMENT_INDUSTRIES_TIRUPUR_INDIA 15/06/2022 15/06/2022

Paunonen, S., Kamppuri, T., Katajainen, L., Hohenthal, C., Heikkilä, P. and Harlin, A. (2019). Environmental impact of cellulose carbamate fibers from chemically recycled cotton. *Journal of Cleaner Production*, 222: 871-881 (online) https://doi.org/10.1016/j.jclepro.2019.03.063

Paszun, D. and Spychaj, T. (1997). Chemical recycling of poly(ethylene terephthalate). *Industrial & Engineering Chemistry Research*, 36(4): 1373-1383 (online) https://doi.org/10.1021/ie960563c

Pedersen, E.R.G., Gwozdz, W. and Hvass, K.K. (2018). Exploring the relationship between business model innovation, corporate sustainability, and organisational values within the fashion industry. *Journal of Business Ethics*, 149: 267-284 (online) https://doi.org/10.1007/s10551-016-3044-7

Potting, J., Hekkert, M., Worrell, E. and Hanemaaijer, A. (2017). *Circular Economy: Measuring Innovation in Product Chain*. PBL Netherlands Environmental Assessment Agency (online) https://www.pbl.nl/sites/default/files/downloads/pbl-2016-circular-economy-measuring-innovation-in-product-chains-2544.pdf 15/06/2022

Raheem, A., Zainon, N.Z., Hassan, A., Hamid, M.K.A., Samsudin, S.A. and Sabeen, A.H. (2019). Current developments in chemical recycling of post-consumer polyethylene terephthalate wastes for new materials production: A review. *Journal of Cleaner Production*, 225: 1052-1064 (online) https://doi.org/10.1016/j.jclepro.2019.04.019

Saarimäki, E. and Sarsama, P. (2021). *Thermoplastic Processing and Composites*. Finnish-Swedish Textile Circularity Day – Webinar, 20th Jan 2021 (online) https://www.youtube.com/watch?v=-A8AEwz8JEQ 13/06/2022

Sert, E., Yılmaz, E. and Atalay, F.S. (2019). Chemical recycling of polyethlylene terephthalate by glycolysis using deep eutectic solvents. *Journal of Polymers and the Environment*, 27: 2956-2962. https://doi.org/10.1007/s10924-019-01578-w

Vehviläinen, M., Määttänen, M., Asikainen, S., Laine, C., Anghelescu-Hakala, A., Immonen, K. and Harlin, A. (2018). *Utilisation of Cellulose from Blended Textiles*. The 8th Workshop on Cellulose, November 13-14, 2018, Karlstad/Sweden

Vehviläinen, M., Määttänen, M., Grönqvist, S., Harlin, A., Steiner, M. and Kunkel, R. (2020). Sustainable continuous process for cellulosic regenerated fibers. *Chemical Fibers International*, 70(4): 128-130.

Wedin, H., Niit, E., Mansoor, Z.A., Kristinsdottir, A.R., de la Motte, H., Jönsson, C., Östlund, Å. and Lindgren, C. (2018). Preparation of viscose fibres stripped of reactive dyes and wrinkle-free crosslinked cotton textile finish. *Journal of Polymers and the Environment*, 26: 3603-3612 (online) https://doi.org/10.1007/s10924-018-1239-y

PART IV

Data and Technology

Data in Textile Circulation

Sheenam Jain[1], Susanna Horn[2] and Kirsi Niinimäki[1]*

[1] Aalto University, Finland
[2] Finnish Environment Institute, Finland
e-mail: *kirsi.niinimaki@aalto.fi

> ~~ *there's no such thing as sustainability without traceability*
> *and*
> *there's no traceability without data…*~~

Introduction

The textile and apparel industry is of global importance, as it is a source of economic value generation, creates substantial employment opportunities, but at the same time results in severe environmental impacts (e.g. Niinimäki et al., 2020). In Europe alone, the average consumption of textiles is 26 kg per person. As the average clothing use time is shortening, the textile waste stream in all Western countries, but recently also in developing countries, is increasing (European Environment Agency, 2021; EMF, 2017). The textile industry has been identified as a buyer-driven value chain that consists of retailers, branded manufacturers and marketers (Muñoz de Bustillo and Grimshaw, 2016). Even today, the textile and apparel value chain is predominantly characterised by a linear economic model of take-make-consume-waste (Niinimäki et al., 2020; Hanuláková et al., 2021; Provin et al., 2021). The current fast-moving pace at which textile and apparel products are produced and consumed is considered unsustainable (Koep et al., 2021; Jestratijevic and Hillery, 2022), and has an adverse impact on climate change, pollution and biodiversity loss, which is often termed the 'triple planetary crisis' (Ütebay et al., 2020). In order to tackle these adverse effects, the way in which the value chain operates today needs to change, that is, shift from linear to circular. This however, requires a full awareness of system wide impacts; for example, rebound effects or inefficient recycling technologies may overweigh the positive impacts of certain CE solutions.

To enable this shift from a linear textile value chain to a more circular one, industry practitioners and academicians have acknowledged that data availability and traceability is a crucial factor in supporting decision-making (Papú Carrone, 2020). Currently, the understanding of the benefits that can be achieved by data is limited, as is the availability of data across the value chain and pilot cases to showcase the achieved environmental, social and economic gains. Moreover, the vague description and lack of consensus on what data means, where it can be found and what types are available, stretch this complexity even further. In isolation, data are rarely usable in their raw form; they need to be processed into meaningful knowledge or indicators or have a context in order for decision-makers to understand their value and implications (Sivarajah et al., 2017). Pilots and concrete use cases are increasingly required to push the issue further and encourage concrete actions.

This chapter first discusses the shift from linear to circular value chains, with the particular focus on data, its role and use in the circular economy. We aim to present both the challenges and the opportunities that may arise from the use of data for enhancing sustainability at the system level, especially how data can build a better circular economy and circular value chains.

From linear to circular value chains

As the population is growing, natural resources are being exhausted, leading to rapidly increasing global and local environmental impacts. Hence, more attention must be paid to developing and supporting a sustainable global economy. The inclination of governance to reduce reliance on fossil fuel imports, to stimulate regional and rural development, to mitigate climate change, and to promote circularity has led to the 'start' of the transition towards a more circular, just and carbon-neutral economy (Ngan et al., 2019; European Commission | OECD, 2022). The desire for more control and the growing necessity of environmental, social and governance (ESG) programmes have made the circularity supply chains and circularity at large, a strategic corporate objective. The circular economy is based on 'principles of closed loops' in which waste is returned to industry as a valuable raw material (Niinimäki, 2018, p. 18). The circular economy approach also tries to extend the use time of products and materials through different kinds of strategies (e.g. re-use, re-design, design for services, recovery and reverse logistics) (Niinimäki, 2018).

The principles of the circular economy can provide a direction to approach the existing challenges in a sustainable way, while paving the path for moderate future disruptions. Circular value chains help drive a circular economy in which operational and environmental impact is not only considered, but also measured. The shift towards a circular value chain will have an impact on the whole organisation, from the strategic to the operational level. To survive this change, value chain stakeholders need to prepare to act swiftly across all the major areas of the organisation, such as products and service design, manufacturing as well strategy and business model development. Better consideration of sourcing, communication/marketing and logistics is also needed. The key points in this transformation are:

- Textile and apparel products must be **designed** for circularity, as 80% of environmental impacts (Tischner and Charter, 2001, p. 120) and 75% of product costs may be the result of decisions taken at the product design stage (Jeppsson and Sjöberg, 2017). Products (or services) are at the core of companies' strategies and operations, so disregarding their circularity in the shift towards circularity will reduce and undermine the transitional impact of such a change.
- **Manufacturing** must consider circularity at its core, optimising end-to-end operations such as sourcing, the supply chain, production and logistics. Moreover, because circularity requires an efficient ecosystem, an organisation's operating model needs to change and expand to include the key operational elements of the broader ecosystem.
- **Business Models** need to be renewed and enhanced to consider the full lifecycle of a product and service, and with this, to expand beyond the traditional considerations of, for example, a sell or rent model into the broader ecosystem context (e.g. Niinimäki, 2018).

Challenges in the textile and apparel value chain

Becoming circular is not easy, especially for companies and industries that have operated in a strictly linear model, such as the textile and apparel value chain. There will never be an easy or optimal time to make drastic changes that will impact or even completely alter the way a company operates. Nevertheless, the sooner the change is made, the greater the potential benefit for early adopters. When it comes to the textile and apparel value chain, becoming circular requires considering the following issues:

- To ensure more sustainable textile/apparel value chains, the entire lifecycle needs to be taken into account. The lifecycle has many phases in which environmental and social impacts may occur, e.g. design, raw materials, fibre/fabric/garment manufacture, transportation, resale, use, and end-of-life processes.

- Many stakeholders are placing demands on the properties of a garment to reduce the environmental impact and mitigate social and environmental risks (i.e. demands on the full lifecycle). These stakeholders may be regulators, but also consumers, local communities, investors, workers, other value chain actors and NGOs.
- In order to make the lifecycle more sustainable and to answer to different stakeholders' demands, actors need more information about each phase of the lifecycle. Companies can no longer take these issues for granted or leave them unobserved or unverifiable; they need to collect data, that is accurate, up-to-date, validated, context specific and representative and use it proactively (i.e. build new products, select new suppliers, develop new business models that cause less negative sustainability impacts).
- In practice, unsustainability may lead to e.g. increasing carbon emissions, water consumption, microplastics, chemical risks, child labour, forced labour, health and safety risks, waste production, eutrophication, and acidification.

Data for circularity in the value chain

Data are defined as facts, numbers, or information stored in various locations such as invoices, contracts, bills of lading, process control systems, monitoring systems, enterprise resource planning (ERP), and emission control systems (e.g. Jain, 2020). By collecting such data, a business can improve its transparency and visibility. Transparency and visibility are critical not only for ensuring the most economically and ecologically sound choices and processes, but also for ensuring data security throughout the value chain. By utilising data, split-second decisions can be avoided in the absence of adequate support.

Leveraging technology and data plays a critical role in enabling change and accelerating the transition to a circular economy for the textile and apparel value chain. By designing apparel for circularity from the onset and providing stakeholders with essential information at each stage of the product's lifespan, significant environmental, financial, and creative value may be harnessed and re-released to benefit both individuals and the community. A universally acknowledged data standard that defines vital product data for collection and exchange has the ability to facilitate both a reverse value chain and large-scale recyclability in the textile and apparel businesses (Jain, 2020).

Current use of data for circularity

Globally, the utilisation of data is growing at a rapid rate; manufacturing processes are being automated, consumer behaviours are being analysed, and statistics are being developed. Subsequently, many actors already have a plethora of data at their disposal or would be able to obtain it. However, determining whether data are being effectively used to advance the circular economy or product sustainability is more challenging.

Depending on the role of the individual actors in the value chain, they mainly use (a) data from their direct suppliers in the form of certificates, quality or technical indicators, material contents; (b) data from their own operations/manufacturing, e.g. energy, water, chemical efficiency, waste flows, production volumes, health and safety issues; (c) data on reclamations and returns; (d) sales and marketing data; (e) consumer data based on interviews or online behaviour; (f) user data, when clothing is provided as a service or rented (e.g. number of washes can be traced using microchips). In some cases, these datasets may be further processed into LCA-type indicators, such as carbon or water footprints.

A company often sources (from outside the organisation) data for its upstream supplier, but typically only for up to tier 1 suppliers (direct suppliers). Retailers/brands also try to collect data from consumers, but this is often difficult due to a lack of contact points with the consumer after purchase, and for data security reasons. Nevertheless, in terms of using data to improve the entire value chain's circularity, efforts are limited. Hence, data are currently mostly used to enhance individual actors' sustainability efforts, such as to improve their own processes, to impact the design of their products (when the actor is the designer), or to source materials from 'responsible' suppliers. There is little evidence of system-wide, structured data use that would support a large-scale transition to more sustainable lifecycles.

Data, stakeholders and circular strategies

To determine whether a business activity is achieving the goals of the circular economy, or at least is on the right path, business leaders must have access to data that measures the circular economy performance of their products and business operations, in addition to the more traditional metrics used to evaluate the business. However, measuring circular economy performance is a relatively new field and can lead to mistakes and circular economy misinterpretations, and result in well-intentioned incremental tweaks to linear systems rather than the adoption of truly circular business models. Measuring circular economy performance also leads to a situation in which data on previously unmeasured aspects of a business, such as the circularity of water flows, needs to be included.

We need to understand how to achieve a circular economy approach that does not focus on only material recirculation. Circular solutions, such as repair services and new kinds of business models (such as lending or leasing) are critical for waste prevention. At an organisational level, the circular economy should also be incorporated into the strategy, risk assessment, and operational goals, for example. In some cases, without the concrete monitoring of impacts and data, circular solutions may become less sustainable than intended, using significant amounts of energy for recycling, increasing transportation and washing in reusing clothing or in rental services or even rebound effects caused by increased consumption due to passing on old garments. In this context, it would make sense for businesses to assess their maturity on the basis of the circular value they generate vs data sharing, as presented in Figure 1.

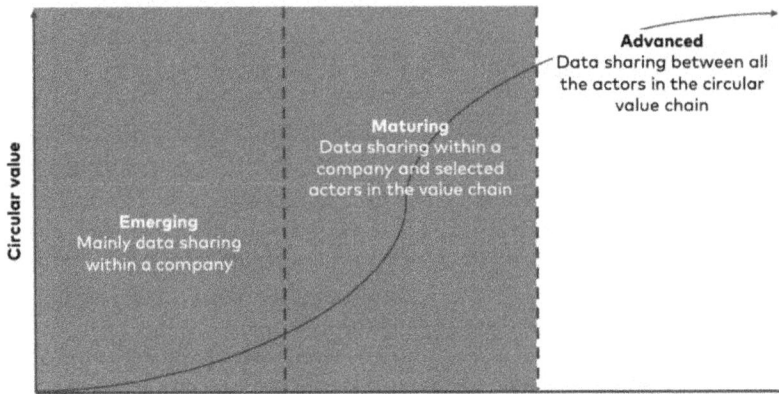

Figure 1. Scale to measure current maturity level of businesses
(Source: Adapted from Nordic Innovation, 2021)

Policymakers also need certain data to develop policies and relevant/corresponding instruments to support circular and sustainable textile markets and phase out the most harmful ones. These data may cover product-specific LCA information to identify the environmental hotspots within a garment's lifecycle or compare various garments with each other. Also, from a policy perspective, national-level material flow data, including imports and exports and information on harmful substances, are highly relevant for redirecting waste flows, planning required capacities or setting restrictions.

The **consumer** also needs certain data for making informed decisions, at the time of purchase the garment and at the end of the garment's useful life. At best, at the moment of purchase, the consumer can be given comparable, verifiable product data about its sources and materials as well as the environmental impacts it has had so far (i.e. LCA data). These environmental data are based on production process data, for example, on the use of energy, raw materials, chemicals and water throughout the cradle-to-gate lifecycle. The consumer also has to choose how to care for the garment and how long to use it for, as well as how to discard it afterwards. For this purpose, the relevant data would contain, for example, maintenance and repair information, washing and drying impacts as well as information about its recycling profile.

Table 1 presents the different types of data that could be collected, shared and applied in different ways in decision-making to enhance the circular economy of the textiles and apparel value chain.

From data to solutions: The road ahead

Assuming that some of the data can be shared between all or selected actors, how do we move from data to solutions? We use the power of algorithms and machine learning. 'The rapid development of new technologies is a game changer for circular business models, and this is opening new opportunities faster than ever before. There are a range of technologies that are transforming value chains and creating well-functioning ecosystems, spanning from artificial intelligence to Internet of Things' (Nordic Innovation 2021). New technical innovations are on the way and are scaling up rapidly. However, using a novel technology is not required or even preferable in all cases, as it requires processing capacity and energy, the hardware may also require the use of rare and critical raw materials. Before implementing new technologies, it is important to assess their game-changing qualities as well as the additional resources their implementation may require.

Moreover, to truly benefit from data use in a circular economy, first, the various actors, the data they create, and how they interact must be understood. Then, the essential skills to realise this potential must be developed. Creating a data-driven culture within firms requires a new mindset, and investing in the relevant technology. Technical and management skills must also be developed so that the new technologies can be integrated into business practices. All these are essential aspects for achieving value in a data-driven circular economy. It is also critical to understand that value can emerge through various channels, and thus be captured through various measures (e.g. Cura et al., 2022).

One of the important developments in this area, especially in the EU, is the development of digital product passports (DPPs) and their related policies. These will electronically register, process and share product-related information between supply chain businesses, authorities and consumers, with an aim to increase transparency across the board. DPPs may contain information that is crucial for circular economy efforts, such as information on durability and reliability, reusability, upgradability, reparability, the possibility of maintenance and refurbishment, the presence of substances of concern, energy and resource efficiency, and recycled content (European Commission, 2022).

Table 1. Data for stakeholders' decision-making

	Design	Fibre and textile production	Garment production	Retail and logistics	Post-consumer processing
Supplier Data	Selecting long-lasting, circular material contents based on supplier data.	Sourcing raw materials from environmentally most sustainable and socially responsible suppliers (including tier X suppliers, not only direct suppliers)	Sourcing fabrics from environmentally most sustainable and socially responsible suppliers (also tier X suppliers, not only direct suppliers)	Sourcing fabrics from environmentally most sustainable and socially responsible suppliers (also tier X suppliers, not only direct suppliers)	
Customer Data	Enabling longer lifecycles by providing long-lasting design. Designing a garment or new product types to fulfil a need, not a trend, by better understanding consumer needs. Enabling easier maintenance and repair by understanding consumers (design for maintenance/design for repair).	Producing only the required amount by understanding demand volumes. Understanding customer (ultimately consumer) requirements and monitoring, collecting and providing the data they demand.	Producing only the required amount by understanding demand volumes. Understanding customer (ultimately consumer) requirements and monitoring, collecting and providing the data they demand.	Optimising logistics and retail volumes (and thus waste) by understanding demand volumes. Restricting/optimising logistics costs/impacts by studying e.g. consumer behaviour related to returns.	Building up required capacity and ensuring steady inflow of post-consumer waste flow by understanding demand volumes and customer behaviour.

Material Data	Selecting long-lasting materials and products. Choosing easily recyclable materials and products. Selecting comfortable materials to wear (impacts longevity).		Selecting the most efficient recycling technology by understanding the technical details and recyclability of various materials.
Company-owned Process Data	Optimising own processes using own process data. Reducing environmental and H&S[1] risks by monitoring own processes.	Optimising own processes using process data. Reducing environmental and H&S risks by monitoring own processes.	Optimising own processes using own process data. Reducing environmental and H&S risks by monitoring own processes.
Product Data			Steering e-o-l[2] garments with certain materials or qualities to correct recycling routes. Acknowledging the accumulation or risks of certain harmful substances.

(Contd.)

Table 1. (*Contd.*)

	Design	Fibre and textile production	Garment production	Retail and logistics	Post-consumer processing
LCA Data	Designing products with least environmental impacts (e.g. reduce carbon/water footprint, eutrophication, acidification, resource depletion, respiratory inorganics). Understanding trade-offs between environmental impacts (e.g. some materials can reduce carbon footprint but increase water footprint).	Becoming aware of fibre and textile production impacts in relation to rest of lifecycle impacts => making strategy for reducing if significant.	Becoming aware of garment production impacts in relation to rest of lifecycle impacts => developing strategy for reducing if significant.	Becoming aware of retail and logistics impacts in relation to rest of lifecycle impacts => developing strategy for reducing if significant.	Becoming aware of post-consumer processing impacts in relation to rest of lifecycle impacts => developing strategy for reducing if significant.
Waste Data	Designing out waste. Designing new products based on waste flows and side streams.	Minimising losses from fibre and textile in production processes by monitoring sources and volumes of losses.	Minimising losses from garment production by monitoring sources and volumes of losses.	Minimising losses from retail by monitoring sources and volumes of losses.	Ensuring steady inflow of post-consumer waste flow for maintaining operations.

[1] H&S; health and safety

[2] e-o-l; end of life

Sustainable supply-chain solutions need global data standards. Throughout the supply chain, consistent information should be available, enabling the use of standardised product identification, such as GS1, Global Trade Item Numbers (GTINs) and Global Location Numbers (GLNs). This could provide the information needed throughout the product journey and throughout the touchpoints in the supply chain. Consistency in how the product is identified through the industrial and logistic phases, and how its movement is communicated throughout the supply chain improves the control of the supply chain and enables control of the product's content and environmental footprint. More reliable information (data standards) and transparency in the supply chain enhances not only its sustainability but also its accountability, agility and how well it is controlled (Cura et al., 2022).

Some value chain actors may be concerned with generating economic value and beating the competition, but others may focus on facilitating a cultural shift, therefore creating value that serves both them and society as a whole. Data-driven solutions can catalyse digitalisation and sustainability to meet this requirement as business models adjust to social and individual needs. To achieve this, we propose the following study directions (based on Pappas et al., 2018):

The role of data actors: By data actors, we mean the stakeholders in the value chain who 'generate the data, own the data, and have the potential to benefit from the data. The data actors are typically involved in top-down approaches of data analysis, but they may also be involved in participatory (bottom-up) endeavours shaping how digital transformation will impact and change society' (Pappas et al., 2018). For digital transformation to succeed, data actors and their activity must be studied more. This needs to be combined with sustainability development and fair and just sustainable societies. FAIR data also need to be studied further (Boeckhout et al., 2018).

Data capacities and availability: More research is needed to understand what kind of regulations are required for data in the digital society (e.g., GDPR). Research should cover the following aspects: 'the capacity, availability and pitfalls of big data, as well as differences between countries, continents, and cultures towards the creation of unified practices and regulations' (Boeckhout et al., 2018). Data collection technologies (e.g. smartphone sensors) and collection methods (e.g. real-time analysis) also need further study from the perspectives of data integration and adoption.

Adoption at the leadership and management level: Our understanding of the adoption and implementation of data-driven methods in different organisations is still insufficient. What kind of leaders and organisational structures are ready to implement these new methods in their decision-making processes and their businesses/industry practices? In particular the value aspect and how it is understood are critical issues. How is value created through transparency in the supply chain? And how can transparency enhance sustainability or social change in the textile and apparel supply chains?

Even though data-driven societal transitions are currently highly sought after and we live in an age of unprecedented access to data, questioning what is really used for decision-making is justified. Access to information may work better in theory than in practice. People and individuals have the capacity to integrate only a certain amount of data into their decision-making, and surprisingly much is left to common beliefs, personal values and, in fact, 'gut feeling' (e.g. Teale et al., 2003). So, the question to ask about data-driven sustainable development is, which data in which format can support decision-making and guide the transition towards more sustainable practices.

Data-driven sustainable development: More information and transparency is demanded of industry and business today. Reliable information is also the foundation of sustainable evaluation of products and their production. The data-driven sustainability approach could solve some of the existing problems in industry and its practices. We need to include data aspects in sustainability development research in order to harness its potential for a more sustainable balance in the textile system. 'Current practices and strategies are expected to build upon data-driven methods, thus we need a deeper understanding of how they can co-exist and co-evolve in the digital society. Various challenges exist before such a transformation can be achieved, and thus we need to change the existing process of how we design information technology and digital practices in our research' (Pappas et al., 2018).

Conclusions

The pervasiveness of data and our capacity to derive insights from them is becoming increasingly rooted in decision-making. In the circular economy transformation, data insights are critical for implementing a successful circular economy strategy. This chapter introduced the broader concept of data in the textile industry and brought it to life by revealing the various potential opportunities and challenges that data-driven circular economy strategies may have.

The continuous development of circularity and connecting this to global data standards will help industries turn liabilities into prospects, which will make products and organisations more productive, efficient or even sufficient. At least data connected to sustainability can provide opportunities to create a better balance in the textile system by longer-lasting products and more accurate production, and by designing closed-loop products that are suitable for recycling. Eventually, supply flows will profit from new resources and the remanufacturing of existing ones, and interrelated supply chain data will make supply chains more flexible and effective, even in times of crisis.

FAIR data (Boeckhout et al., 2018) and effectual product system design are critical for expediting the transition to a circular economy. What challenges are involved in accessing data when moving towards a circular economy and how can we overcome them? What is a well-designed circular economy and how can we ensure it is promoted enlightened by good data? Realising the benefits of data-based circular solutions requires understanding the prospects and challenges of digitalisation. What is the present situation and what else is needed to leverage data at the system level for circular transition? These are only some of the questions that individuals, organisations or economies need to address in order to begin their circular journey.

References

Boeckhout, M., Zielhuis, G.A. and Bredenoord, A.L. (2018). The FAIR guiding principles for data stewardship: Fair enough? *European Journal of Human Genetics*, 26(7): 931-936. doi: 10.1038/s41431-018-0160-0.

Cura, K., Jain, S. and Niinimäki, K. (2022). Transparency and traceability in the textile value chain. Report, Aalto University, Finland. https://aaltodoc.aalto.fi/handle/123456789/117564

EMF Ellen MacArthur Foundation (2017). *A New Textiles Economy: Redesigning Fashion's Future.* https://www.ellenmacarthurfoundation.org/assets/downloads/publications/A-New-Textiles-Economy_Full-Report.pdf

European Commission | OECD (2022). *Policy Brief on Making the most of the Social Economy's Contribution to the Circular Economy.* doi: 10.1787/20794797.

European Environment Agency (2021). *Textiles in Europe's Circular Economy.* https://www.eea.europa.eu/publications/textiles-in-europes-circular-economy

Hanuláková, E., Daňo, F. and Kukura, M. (2021). Transition of business companies to circular economy in Slovakia. *Entrepreneurship and Sustainability Issues*, 9(1). doi: 10.9770/jesi.2021.9.1(12).

Jain, S. (2020). *Big Data Management Using Artificial Intelligence in the Apparel Supply Chain: Opportunities and Challenges.* Doctoral dissertation. Department of Textile Management, University of Borås, Sweden.

Jeppsson, J. and Sjöberg, J. (2017). *Establishing a Cost Model when Estimating Product Cost in Early Design Phases.* Master's thesis. Blekinge Institute of Technology. Available at: https://www.diva-portal.org/smash/record.jsf?pid=diva2:1136656 (Accessed: 6 June 2022).

Jestratijevic, I. and Hillery, J.L. (2022). Measuring the "Clothing Mountain": Action research and sustainability pedagogy to reframe (un)sustainable clothing consumption in the classroom. *Clothing and Textiles Research Journal*. doi: 10.1177/0887302X221084375.

Koep, L., Morris, J., Dembskin, N. and Guenther, E. (2021). Buying practices in the textile and fashion industry: Past, present and future. *In:* Matthes, A. et al. (Eds.), *Sustainable Textile and Fashion Value Chains*. pp. 55-73. Cham: Springer. doi: 10.1007/978-3-030-22018-1_5.

Muñoz de Bustillo, R. and Grimshaw, D. (2016). *Global Comparative Study on Wage Fixing Institutions and their Impact in Major Garment Producing Countries*. Available at: https://www.researchgate.net/publication/317692572_Global_comparative_study_on_wage_fixing_institutions_and_their_impact_in_major_garment_producing_countries (Accessed: 6 June 2022).

Ngan, S.L., Howb, B.S., Tengc, S.Y., Promentillad, M.A.B., Yatime, P., Erf, A.C. and Lama, H.L. (2019). Prioritization of sustainability indicators for promoting the circular economy: The case of developing countries. *Renewable and Sustainable Energy Reviews*, 111: 314-331. Available at: https://www.sciencedirect.com/science/article/pii/S1364032119303077?casa_token=_DrNdbiQDiEAAAAA:cffWsey 6MaXtjK_ 0Ekmc8D8qTs5PZFjX6F MGwHyh AMNNN1MZaCPnAcLI4X_ziauPJXqo4T_ oumo (Accessed: 6 June 2022).

Niinimäki, K. (Ed.) (2018). *Sustainable Fashion in a Circular Economy*. Aalto ARTS Books. https://aaltodoc.aalto.fi/handle/123456789/36608

Niinimäki, K., Peters, G., Dahlbo, H., Perry, P., Rissanen, T. and Gwilt, A. (2020). The environmental price of fast fashion. *Nature Reviews Earth and Environment*, 189-200. doi: 10.1038/s43017-020-0039-9.

Nordic Innovation (2021). *Data Sharing for a Circular Economy in the Nordics*. Available at: https://www.nordicinnovation.org/CEdatasharing (Accessed: 5 June 2022).

Pappas, I.O., Mikalef, P., Giannakos, M.N., Krogstie, J. and Lekakos, G. (2018). Big data and business analytics ecosystems: Paving the way towards digital transformation and sustainable societies. *Inf. Syst. E-Bus Manage.*, 16: 479-491. Available https://doi.org/10.1007/s10257-018-0377-z

Papú Carrone, N. (2020). Traceability and transparency: A way forward for SDG 12 in the textile and clothing industry. *In:* Gardetti, M. and Muthu, S. (Eds.), *The UN Sustainable Development Goals for the Textile and Fashion Industry*. pp. 1-19. Singapore: Springer. doi: 10.1007/978-981-13-8787-6_1.

Provin, A.P., de Aguiar Dutra, A.R., de Sousa, I.C.A., Gouveia, S. and Cubas, A.L.V. (2021). Circular economy for fashion industry: Use of waste from the food industry for the production of biotextiles. *Technological Forecasting and Social Change*, 169: 120858. doi: 10.1016/j.techfore.2021.120858.

Sivarajah, U., Kamal, M.M., Irani, Z. and Weerakkody, V. (2017). Critical analysis of Big Data challenges and analytical methods. *Journal of Business Research*, 70: 263-286. doi: 10.1016/j.jbusres.2016.08.001.

Teale, M. (Ed.) (2003). *Management Decision-making: Towards an Integrated Approach*. Harlow, UK: Pearson Education.

Tischner, U. and Charter, M. (2001). Sustainable product design. *In:* M. Charter and U. Tischner (Eds.), *Sustainable Solutions: Developing Products and Services for the Future*. pp. 118-139. Sheffield, UK: Greenleaf Publishing.

Ütebay, B., Çelik, P. and Çay, A. (2020). Textile wastes: Status and perspectives. *In:* Körlü, A. (Ed.), *Waste in Textile and Leather Sectors*. pp. 39-56. London: IntechOpen. doi: 10.5772/intechopen.90014

CHAPTER
11

Policy Paired with Technology

Sajida Gordon

Nottingham Trent University, UK
e-mail: sajida.gordon@ntu.ac.uk

Introduction

Adverse impacts from the fashion industry, such as carbon emissions, water pollution and exploitative labour practises are clearly visible on the planet and to humanity. Manufacturing activities, the overuse of resources and consumption have contributed a great deal to textile waste and negative impacts on the environment. There is clear indication that both upstream and downstream activities are contributing to greenhouse gas (GHG) emissions; with the largest emphasis coming from upstream activities, such as material input and processing, these contribute 70% to GHG emissions (McKinsey & Co, 2020). Reducing emissions and textiles waste is not dependent on a singular element to change these statistics, but rather needs a multifaceted approach and collaborative cohesion. However, a key enabler such as legislation, has the power to allow for tangible, sustainable and circular change. The EU strategy for sustainable and circular textiles emphasises the need for better usage of textile and extending lifetimes, through mandatory Extended Producer Responsibility (EPR), digital product passports (DPP) and eco-design requirements (European Commission, 2022b). Furthermore, the emphasis on re-use and repair business models are also stipulated as an objective to further drive the longevity of textile products. The UK Waste Prevention Programme (WPP) essentially takes a similar direction with the intention of preventing waste; while extending the capacity of textile lifetimes given that a recent report stated:

> There is significant scope to improve the durability, reparability, recyclability, and use of recycled content in garments. Extending the active life of 50% of the clothing purchased in the UK by nine

months could lead to an 8% reduction in the total carbon footprint and 10% reduction in the total water footprint of clothing in the UK.
(UK Department of Environment, Food and Rural Affairs, 2021)

Furthermore, evidenced by the EU strategy for sustainable and circular textiles, there is an emphasis on incorporating technology, such as product passports, materials databases, and corporate reporting. As the requirement of, reporting on emissions, supply chain transparency and waste tracking, is becoming a necessity, it's apparent that technology will be embedded in multiple systems throughout supply chain and the post-consumer phase. This suggests technology is becoming a vital tool to enable a successful circular and sustainable textiles industry. It can be said that the processes of making clothing has evolved throughout time, where there has been a greater movement towards sustainability in fashion. However, progress on a complete circular system still has a long way to go. A garment goes through various stages from raw material extraction to textile milling, garment production, consumption, and disposal. Within each of these stages, the processes present their own complexities and require further tasks prior to arriving at a complete garment; before it is worn and discarded at the end of its life. To manage these processes and maintain extending the life of textile products, requires a solution that uses digital accuracy, speed of information, and interconnectivity (CB insights, 2022). The assistance of technology together with the force of legislation, can enable the textiles industry to achieve its full potential of circulating textiles and reducing waste. Participation is required from both consumers and producers, without both actively and drastically making equal changes, the negative impact effecting the environment will not shift.

This chapter looks at both policy regions in the EU and UK, addressing both pre- and post-consumer stages in identifying technologies; and how they support the management of textile product lifecycles to achieve complete circularity. There will also be a focus on a government-supported initiative that is assisting in implementing new potential legislative measures, in addition, businesses are engaging with the initiative in adopting circular practises; such as collecting data to identifying their carbon and water footprints of current textiles products. This becomes a starting point to establish a baseline for improvement actions. As businesses start implementing activities to enhance circular models, the data reported, will make it more clearly visible concerning how much carbon, water and waste impacts are reduced. However, there are limitations in accuracy of this data, as reliance is on third parties and external suppliers to retrieve the necessary information, which may be biased. This chapter also focuses on the importance of digital connectivity that supports a globalized supply chain network in monitoring, visibility, and data sharing; and in eliminating some of the obstacles that are currently faced in the fashion and textiles supply chain. Digital product passports have been introduced through the EU strategy for sustainable and circular textiles, capturing information about a garment's journey from raw material through to various second and third life options, such as re-sale and reuse models. A final consideration is given to end of life in terms of how technology and innovation are providing solutions in creating input materials from textiles waste. Challenges here are, the need for scalability of these technologies and the facilities to accommodate constant waste streams. A further challenge is waste material becoming feedstock for production, enabling recycling to become an integral step in the product development stages and not only at end of life. Capturing both manufacturing and consumer waste could disrupt the current model of take, make and waste. It will demonstrate that waste associated with value changes the dynamics of our interaction with material objects.

Circular initiatives

WRAP Textiles 2030

Initiatives are needed to facilitate the reduction of environmental impact through support and the development of a circular model. A prominent group, The Waste and Resources Action Plan (WRAP) a not-for-profit organisation (NGO) established in the UK, delivers such initiatives. WRAP operates across industries and extends to further regions of the world, tackling 'the causes of the climate crisis' (WRAP n.d), working with food, plastics, and textiles in achieving net zero. Their current initiative, Textiles 2030, focuses more specifically on the fashion and textiles sector. This initiative 'aims to engage the majority of UK fashion and textiles organisations in collaborative climate action...on fast-tracking the UK Circular Economy' (WRAP n.d). Textiles 2030 has been gaining significant attention. It is supported by the UK government, Department for Environmental, Farming and Rural Affairs (DEFRA) and voluntary agreements from numerous businesses, to act on preventing waste and reducing emissions. Textiles 2030 has followed on from the Sustainability Clothing Action Plan (SCAP) 2020, a preliminary initiative conducted by WRAP, that brought together fashion and textile industry businesses, where achievements were made in reducing water usage and meeting carbon targets. Textile 2030 has developed 3 categories that aim to work together, these are circularity, metrics and implementing policy changes. The objective is to 'cut carbon by 50%...consistent with limiting global warming to 1.5°C, in line with the Paris Agreement on climate change' (UKFT 2021).

Figure 1 demonstrates the intended pathway to achieving circularity and reducing waste. WRAP anticipates driving change by designing

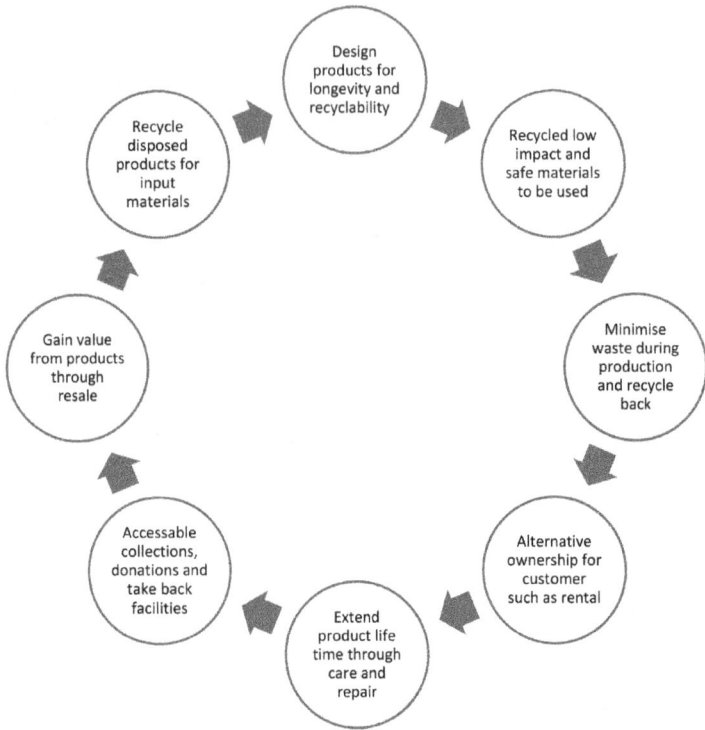

Figure 1. Textiles 2030 circularity pathway (WRAP, 2021)

out waste, closing the loop and working with multiple UK businesses to replace the current linear model to become circular. Policy will play a crucial role in this plan, such as implementing the Extended Producer Responsibility (EPR), stringent standardised requirements for eco-design and durability, and fiscal instruments. WRAP aims to support DEFRA to establish these new potential policies, through engagement with businesses, in identifying which measures are the most appropriate.

Textiles 2030 Metrics

Metrics is the tool that quantifies the task of achieving circularity. To achieve this, a footprint tool is developed by WRAP where businesses, such as brands, retailers, reuse and recycle organisations, can input their water, waste and GHG emission data in relation to their activities. The specific data that businesses are required to put into this tool, are the weight of products and fibre compositions. The products are analysed by Stock Keeping Unit (SKU) level, which indicates the differentiation of price, colour, manufacturer and size for each style and product. (Agarwal 2014). Additionally, for products that are sold in multi-packs, data would need to be multiplied by the number of items within each pack. For larger fast fashion brands all of this equates to copious amounts of data, as the units sold are at a greater volume. A barrier which many businesses may face is the lack of technical knowledge and infrastructure to accommodate data acquisition. However, this task of reporting such data allows businesses to systemise and establish the environmental impacts; while slowly integrating data monitoring to become part of their product development process, enabling conscious planning of sourcing by meeting eco-design requirements. Furthermore, data acquisition in this way is permitting closer working relationships amongst retailers and manufacturers, as much of this data retrieval is required to be accurately communicated between supply base and retailers. This data serves as baseline data, from here actions are considered in terms of how businesses can improve their carbon footprint. There are also specific requests to identify the recycled and organic materials used, these need to be evidenced with certifications. Brands and retailers are also expected to state actions around fabric dying and other processes that contribute to lower environmental impact. As suggested by the Corporate Value Chain Accounting-Reporting Standards, utilising global averages is far better than trusting inaccurate data (GHG Protocol, 2004). Therefore, WRAP makes use of the global trade data of materials flow: for example, identifying cotton's environmental impacts, WRAP would establish where the fibre is likely sourced from; how it is prepared for spinning; where is it made into fabric; how and where is it dyed and what kind of transportation was used between these processes. Furthermore, fibre loss is also accounted for as this tends to be a factor when developing textiles, such as the input of cotton lint is less compared to the output of the cotton fibre. Each stage has a determined impact factor for fibre type and country location it is sourced from or manufactured. Once this data is analysed, together with the data provided from businesses, this ensures relative accuracy. Repair and rental models also contribute with their data, as these are secondary products, the data collected are the weight of quantities of garments, that are being processed through these services. A segment of the footprint tool has been developed so that recycling and reuse businesses can also report. The data consists of the volume inbound and outbound, and the location of where textiles products are going to. The relevant environmental impact is calculated throughout all these processes (Gray, 2022). Collecting data in this way, by showing the environmental impacts of current activities, and developing improvement actions, would enable textile products to be designed using eco-friendly methods for effective post-consumer options. The three categories of Textiles 2030: circularity, metrics and policy implementation, are a clear formula for shaping a product's lifetime and managing its extended potential.

The Importance data acquisition

The UK government has committed 'to legislating for Smart data...to provide consumers and small businesses with the power to enable trusted third parties to help them access, make sense of, and use their data' (UK Department for Digital, Culture, Media and Sport, 2022). This strategy allows the drive for collaborative partnerships across industries, which could imply that third parties will enable businesses to use and optimise data. Furthermore, the UK's Waste Prevention Programme stipulates that the essential need of data acquisition is a fundamental requirement, to ensure the textiles industry is efficiently utilising and has access to 'secondary materials as input' (UK Department of Environment, Food and Rural Affairs, 2021). A process referred to as Industrial symbiosis (IS) essentially underpins 'waste utilization in a productive chain, which finds ways to use the waste from one industry as inputs or raw materials for the other, all supported by business collaboration' (Boom Cárcamo and Peñabaena-Niebles, 2022 p.2). This beneficial strategy can positively impact the environment, economics and social aspects and adds value to waste. However, to enable a successful IS, digital connectivity of waste is crucial, such as embedded codes that can be tracked and are accessible to businesses. Lack of this accessible technology for businesses indicates a challenge in identifying where material outputs are, leading to non-participation of IS. Data inputs and reporting, such as what is being done with Textiles 2030, enables suppliers and manufacturers to develop best practises for environmental needs. This becomes a crucial factor in regions where sustainability in textiles is not a legislative measure, moreover, this is further secured, made accurate and monitored when used along with transparency platforms such as blockchain (Tripathi et al., 2021).

Supply Chain

Supply chain obstacles

Adopting a circular economy and managing sustainability is underpinned by dealing with the intricate and individual processes of textile products, which can achieve an absolute closed loop system; that extends product life. This would mean the supply chain has to come under scrutiny and cannot be overlooked. Governments are gradually identifying the need to disrupt the linear model of producing waste and unsustainable practises. Together with supply chain participants, they must comprehend more about the interrelationships between micro-obstacles within the supply chain, to then obtain a greater realization of the critical macro-obstacles that impede the adoption of a circular economy (Chen and Lin, 2021 p. 4). Examples of this could be the lack of alignment principles in the supply chain or the absence of different information exchange systems between stakeholders.

Digital supply chain

Very few leading supply chain executives implement programs for risk management and compliance into their business models and technology systems to monitor risk. Those that do, a driving factor for this is likely, disruptive events that could impact revenue or company reputation. However, to adopt a circular system and effectively manage the lifetime of textile products, manufacturers together with brands and retailers need to incorporate compliance through each value chain; in ensuring that necessary actions to enhance a circular economy are being taken. This can be achieved by the utilisation of technology to monitor and capture data consistently. Nevertheless, this does pose a challenging task for many participants in the supply chain, specifically in emerging or developing countries, where there is lack of infrastructure, no governmental incentivisation or regulations and organisational silos. When deciphering the supply chain, it is evident that during each stage there needs to be 'collaborative decision making' (Butner, 2010 p.25), with the need for compliance monitoring adopted at micro levels, such as sourcing, manufacturing, and logistics. As these actions are part of a globalization network, it may seem, that technology demonstrates an absolute solution of visibility and governance. Transparency and visibility have been very much of a focal point in the fashion industry, the driving factors stem from organisations demonstrating sustainable and ethical practices to the wider audience, to prove the worth of their sustainable pledges. In contrast, as directed by the EU strategy for sustainable and circular textiles, the tools for transparency are becoming an imperative system for managing textiles, tracking life journeys and processes to account for where and how textiles are located and utilized. Looking closer to the UK's newly published digital strategy in June 2022, the government commits to incentivising 'innovative businesses in the technology sector and support cutting edge computational R&D' with fiscal benefits to cover 'cloud computing and data acquisition' (UK Department for Digital, Culture, Media and Sport, 2022). IBM (International Business Machines) specialises in producing computer hardware and software, as well as offering services of cloud computing and data analytics. IBM's involvement in research and innovative development has also demonstrated significant inventions within the technology sector. Their development of a cognitive supply chain for various industries has been present for the last decade. One study observed, 'This technology has enormous potential in improving the information flows of supply chains' (Moretto and Macchion, 2022 p.2). IBM's smart supply chain utilises data to provide traceability, advanced analytics for decision making and monitoring with a connected supply chain. This integrated cloud system is amalgamated with Red Hat, an open-source systems provider, allowing for scaled-up data solutions. IBM has been able to expand into supply chain activities using technology such as Blockchain, IoT, smart sensors and track and trace (Smith 2021, p. 3).

Red Hat provides a cloud-based infrastructure where higher volume data is collated and assimilated. As the company states:

> *'Red Hat delivers hardened, open-source solutions that make it easier for enterprises to work across platforms and environments, from the core datacentre to the network edge. By operating transparently and responsibly.* (Red Hat n.d)

IBM aims to meet the government's strategy on digital enhancement for the fashion industry, working with the UK Fashion and Textiles Association (UKFT), Tech Data and the Future Fashion Factory to deploy a system, which makes the supply chain transparent. IBM's development of a cognitive and smart supply chain was underpinned by three integral principles, 'instrumented, interconnected, and intelligent' (Butner 2010, p.26). These Fundamental concepts are explored further with IBM piloting a 'supply chain optimisation platform' with a consortium of UK fashion retailers (Bennett, 2021). The three principles demonstrate the benefits for a digital supply chain.

Instrumented – Smart technology provides a system of regular monitoring and ensuring supply chains are running efficiently, and sustainability activities are being met benefiting from 'preventative maintenance' (Butner 2010, p. 60).

Interconnected – Data sharing, visibility of actions and decision-making are achieved through the solution of interconnectivity. This allows for improvements to be consistent within the product engineering, sourcing and adoption of activities that enhance a circular economy.

Intelligent – A constant evolution of innovative product development through analytics meeting the needs of circular economy. Ensuring sustainability and circular processes are applied and thought of throughout the lifetime of a textile product (Demestichas, and Daskalakis, 2020).

Transparency

Digital product passport

A Green Deal plan released by The European Commission in 2022 aims to legislate 'sustainable products to be the norm in the EU' (European Commission, 2022a). This strategy attempts to 'tackle fast fashion' waste and enable circular textiles, and has introduced digital product passports, which aims to address and facilitate repair, recycling and tracking of materials. This system would also ensure consumer involvement by making them aware of the recyclability options and how environmental friendly the products they purchase are (Waldern., Stienbrecher and, Marinkovic 2021). Additionally, in the UK a comparable echo, although not substantiated yet, seems to suggest a similar strategy. The UK Waste Prevention Programme (WPP) specifies the beneficial use of introducing product passports, such as 'improving access to data and enable more efficient use of resources' (UK Department of Environment, Food and Rural Affairs, 2021), enabling efficient recovery and reuse. Digital product passport (DPP) technology is centred around blockchain, a tag or digital code, which stores all information of textile products and their lifecycle journey, from start through to re-use and recycling phases (Waldern, Stienbrecher and Marinkovic, 2021). As stated in WPP, identifying waste and its location is also a key factor (UK Department of Environment, Food and Rural Affairs, 2021). An electronic waste tracking system is necessary to provide granular level data to ensure the correct management of waste, as well as to identify how waste is used as input material. Current traceability and visibility of data has been extremely limited to the sourcing nature of products, by providing information on the environmental impact in manufacturing, or by using QR codes for consumers to be able to identify a product's sustainable story (Williams, and Hodges, 2022). However, it can be argued that the use of QR codes is a marketing strategy for businesses, rather than managing the material flow. Creating a comprehensive and interactive data-driven system, which is inclusive of re-use and recycling phases, allows all participants to monitor and track a garment's journey; ensuring second or third life and correct disposal is achieved. This gives a more meaningful purpose and true sense of accountability for everyone involved in the entire life cycle of a garment.

The actual lifecycle of a garment contains a larger number of stakeholders such as brands, designers, material producers, garment producers, logistics services, packaging providers, online and brick-and-mortar retailers, consumers, C2C and second-hand marketplaces, repair and refashion services and recycling operators. (Järvinen et al., 2021. p. 6)

This view suggests that all parties involved with the handling of textiles products, must embrace the concept of data collection and assimilation for the purpose of circularity. Accessibility and transparency need to be acquired by all involved, including market surveillance and public authorities, to refine governing obligations for eco design requirements and authenticating the accuracy of information provided (European Commission, 2022a).

DPP as blockchain technology, provides added benefits of trust, security, and immutability, which is essential for this approach to work efficiently. A policy that allows for the use of DPP technology should also aim to achieve the phasing out of unsustainable products. Equally, a built-in communication system would allow consumers, repair services, recycling units and second-hand retailers to provide information on the performance and appearance of a garment. This in turn allows manufacturers and designers to understand what improvements need to be made (European Commission, 2022a).

The food sector demonstrates the use of DPP or better known as QR codes. This technology provides 'an information system that effectively tracks and monitors the entire food process from farm to fork, as well as critical stages of production, logistics, and land auditing' (Pandey et al., 2022, p. 7). In addition, the food supply chain and agricultural supply chain are combined and viewed together, as the monitoring of 'food quality, food safety, food wastage, food price volatility, deforestation, and carbon emissions' (Pandey et al., 2022 p. 1), are also vital. This accessibility for consumers ensures food is of high quality and fit for human consumption, allowing consumers to trace ingredients and origins of products. Email and SMS QR codes provided on food packaging allows simple communication access for customer service. Furthermore, consumers can provide feedback on the product to businesses. (Menon, 2022). A successful DPP system enabling a circular system and prolonging the life of garments, will work well alongside stringent sustainable performance requirements. Here we develop a complete change by using a cloud-based lifetime management system for textiles, which can expand to other sectors (Ertz et al. 2022). The EU and UK plans to legislate a strategy for DPP that encompasses and ensures that the complete journey is geared to close the loop. It could further expand the concept of the Extended Producer Responsibility scheme (EPR) in propagating and pressing responsibility onto consumers, recyclers, reuse and repair services (Eco-Standard Org, 2022). Each product begins to be viewed in a more critical way by all those that come into to contact with it, allowing environmental consequences to be understood and the need to reduce waste. The Circular Product Data Protocol developed by the Circular ID initiative, underpins the importance of comprehensive data retrieval with all partners and value chains (Circular ID initiative, 2021). Furthermore, harnessing DPP technology enables the development of digital twins, a digital replica of the physical product; and the use of real-time records of status and lifetime data. This would make a product which is essentially invisible, more visible to all (Scheiner, 2022).

End of life

Post-consumer waste

The EU's Circular Economy Action Plan addresses the entire linear model from extraction to disposal. However, while this policy enforces the development of new systems within the pre-consumer stages, there is still an imminent necessity to address textile products currently present at the opposite end. With a greater emphasis on the sustainable product framework, eco-design requirements would mean that 'new products need to fit into circularity' (European Commission, 2020). This demonstrates how crucial recycling facilities and systems are for products at their end of life, allowing waste to be viewed as valued material inputs. Enabling technology has been trending rapidly in many areas, and within the system of recycling technological application is vital in 'boosting the sorting, re-use and recycling of textiles...through innovation' (European Commission 2020). As the action plan calls for innovation, industry stakeholders must consider more investments into technology, as these methods seem to be a solution to aid a more sustainable fashion landscape. When we look at the factors involved in recycling, we can identify improvements in collection and sorting that are needed before recycling can proceed.

Digitally enhanced collections

Firstly, when we consider the collection of textiles waste, a clear implemented system is needed for the public to donate or dispose of clothing in the correct way. Collection and sorting practices involve quite complex tasks. Information systems are necessary to ensure the accuracy of recording and reporting metrics of how much is processed and at what rate. However, the lack of visible connections between textiles waste streams processes means that data integration is lacking. The planning and efficiency of transportation is also a factor to consider here, as reducing carbon emissions, through transport optimisation, would also demand strategic planning. Companies such as Sensoneo, catering for waste collection and management, aim to develop digital solutions that 'digitize workflows encouraging automation & optimization' (Sensoneo, n.d). The digital solution would require all actors to be responsible and involved in contributing to textiles waste streams, while being in alignment with the EPR. This suggests the management of EPR can also benefit from end-to-end software solutions, through tracking and digital communication, the producer can ensure their textile products are appropriately recycled. Table 1 shows Sensoneo's digitized solutions within the collection phase of recycling. They work with various actors in the post-consumer phase, such as take back schemes and processing facilities. The digital solutions provide multiple benefits demonstrated in Table 1, fulfilling the requirements of EPR and enabling waste optimisation.

Table 1. Digital solution and benefits for waste management (Senconeo, n.d)

Stakeholders/ Actors	Tech / Digital solutions	Benefits
Take back systems	Transparent workflow Multiple entry points for data inputs. Smart tools for stakeholders' data response Automation of reporting	Complete digitalised platforms, reduction in workloads and inefficiencies, Helps target EPR, PRO, DRS
Waste monitoring	IoT networks monitoring of collection bins to collate data Ultrasonic smart sensors deployed in bins Citizen App for reporting, monitoring, and location finder Smart waste management system – cloud-based platform for data driven operations.	Managing waste smarter and efficiently, continuous visibility and real-time response, remote managements reducing less administrative workload
Logistics	Web portal for operators to create orders, create collection routes and assign drivers Driver accessible app Algorithms for assignment of loads and route planning.	Transparent invoicing streamlining pickups and discharges, more profit per drive.
Collections	Smart mobile application to place pick up requests Smart sensors: Automation of pick-up placement when bins are full Smart sensors service verifications	Fully automated workflow creates speed and accuracies, Transparent waste stream, Lower collection cost, user friendly technology
Processing facilities	Smart applications for data inputs for reporting Visibility on all deliveries and assigned waste Mobile application and web portal accessibility	Reliable and real time data, Workflow is processed in a speedy manner,

Analysing Table 1 clearly demonstrates the need for digital solutions and showing how this benefits the smooth synergy and flow of textiles needed for appropriate collection and management. Other businesses such as Evreka, also provide software and hardware solutions for smart waste management. They work together with Producer Responsibility organisations (PRO) in ensuring 'that the entire EPR process is managed following government regulations… Evreka's all-in-one EPR solution facilitates every little detail about the chain of custody' (Evreka, n.d). The Material Recovery Facility record and analyse data of material inflow. This allows Evreka to optimise on decision making ensuring correct materials are sorted and managed to reach the appropriate end-buyers with real-time data.

The textile collection phase is the initial step of post-consumer recycling, and therefore, it needs to be conducted and implemented efficiently and effectively to meet the demands of developing a circular model. Here we can begin to establish the start of a re-use and recycle journey, which is vital when we consider the lifecycle of a garment and how to manage it sustainably.

Sorting with Near-infrared Spectroscopy

Recycling also relies upon recognition of fibres and sorting into appropriate categories. Typically, the process of sorting discarded garments and textiles has been a manual process, which is both costly and labour intensive (Sandvik, and Stubb 2019). Workers within sorting facilities have relied on clothing labelling to determine fibre composition. Studies suggests that 'up to 41% of labels on blended materials contain inaccurate information' (Cura et al., 2021 p. 1). Furthermore, 'many synthetic fibres are finished to possess similar surface parameters, which confuse the identification process' (Zhou et al., 2019b, p. 201). Clearly, manual sorting methods are not reliable in constituting and guaranteeing an effective recycling process. Various other methods are used, such as ISO 1833-1 Dissolution Behaviour, Microscopy Detection, DNA recognition and Thermal Behaviour Detection. Yet, these methods, although accurate, are time consuming and input materials need to be prepared usually requiring lab-based environments (Cura et al., 2021). Raman Spectroscopy and Near-infrared Spectroscopy (NIRS) have also been used to determine the composition of textiles and classifications; and specifically, NIRS has gained momentum for fibre recognition in the textile industry (Zhou et al., 2019b).

> *NIR spectroscopy is based on molecular absorptions measured in the near infrared part of the spectrum. The infrared light from a light source is partially and selectively absorbed by the target surface, and the reflected light creates a characteristic spectrum of each fibre type or blend combination.* (Tossavainen, 2019)

The NIR spectroscopy method is beneficial as materials are not required to be prepared or pre-treated (Zhou et al., 2019a). This is essential when working through larger quantities of textile waste. Smart fibre sorting companies, such as Fibresort, use NIRS technology and can 'sort 900 kgs of post-consumer textiles per 1 hour' (Weiland Textiles, n.d). The speed and volume of this technology is clearly more efficient than manual sorting. The established Textiles for Textiles R&D project which was launched in 2009 aimed to develop an industrial scale machine that automated the sorting of textile waste (Smart fibre sorting n.d). By collaborating and joining forces with Weiland Textiles and Valvan Baling Systems, Fibresort was developed in 2018 as a proven NIRS technology for sorting of post-consumer waste. Furthermore, to expand innovation, a consortium of experts was formed together under the EU funded project 'Interreg-project Fibersort: Closing the loop in the textiles industry', which consisted of Weiland Textiles, Valvan Baling Systems, Circular Economy, Prototex Cooperation, Worn Again Technologies, Smart Fiber Sorting and Leger des Heils ReShare (Smart Fibre Sort, n.da).

Table 2. Interreg-project Fibersort consortium (Weiland Textiles, n.d)

Company	What does it do?
Weiland textiles (*Netherlands*)	*Purchase and Sorting of collected textiles, Marketing and Sales of sorted textiles, Innovation – Fibersort*
Valvan Baling systems (*Belgium*)	*Preparing, feeding, and baling machinery manufacturer*
Circular economy (*Netherlands*)	*Providing metric measurements of the circular economy and overviews of material flow.*
Prototex Cooperation (*Belgium*)	*Recycles natural, synthetic, and technical textiles waste*
Worn again technologies (*United Kingdom*)	*Utilizes Polymer recycling technologies*
Smart fibre sort (*Netherlands*)	*NIR Spectroscopy fibre sorting*
Leger des Heils ReShare (Salvation Army ReShare) (*Netherlands*)	*Collect used clothing and textiles, to be given away to those in need or to be sold in ReShare stores.*

This consortium of experts worked together to establish a systematic and coherent process of collection, sorting, and recycling, utilizing robotic technologies, machinery, and metrics with further improvements and innovation. (Smart Fibre Sort, n.d). To further expand on recycling post-consumer waste, businesses need to tackle the textiles waste in manufacturing; tackling waste material during manufacturing could constitute as mass production of recycling feedstock. To achieve this, manufacturers and producers could establish more appropriate and accurate components of garments that make recycling simpler by incorporating eco-design principles. Weiland Textiles is developing a new program called Trim Clean Technology that aims to tackle the issue of trims and components of waste garments (Smart Fibre Sort, n.d). The UK Circular Economy Package (CEP) demonstrates that, the duration of how long a product lasts is determined by its repeated use.

> *Shifting towards a more circular economy will mean we optimise our use of resources within the economy by increasing the duration of a product's useful life and ensuring when a product has reached the end of its life, its resources can be productively used repeatedly, so creating further value.*
> (UK Department of Environment, Food and Rural Affairs, DEFRA, 2020)

Here we can understand the significant need to establish closed loop systems, whereby the lifetime management of textiles is controlled and achieved. This could assure that all waste materials are recycled into new outputs; and businesses are not reliant on virgin raw materials. This change in the manufacturing and sourcing paradigm requires the imminent need for participation of various actors, such as the recyclers and manufacturers. This approach could highly benefit the EPR scheme by enabling it to work better where producers are actively collaborating with partners across the complete chain.

Recycling dyes

A problematic process with recycling textiles has been the need to deal with chemical finishes and dyes. The wide commercial use of synthetic chemical colourants within textile production poses harmful impacts, not only to humans but also on the environment. Additionally, the application of some colourants require adhesion with the help of some substances at molecular level, which recycling textiles mechanically cannot remove. The method of chemical recycling allows the removal of chemicals and dyes; however, it contributes to the issue of substance leaching. While the removal of these substances can be achieved, it can compromise the integrity of the material being recycled as 'It is speculated that technical issues may arise, such as decreased dyeability or the need for an additional purification process step' (Le, 2018, p. 44).

Advances in technology are presently making an appearance to successfully tackle recycling, without dye substances impeding the long-term process of re-engineered materials. One such company, DyeRecycle, not only recycles fibres, but also recycles dyestuffs. The technology involved utilises non-hazardous liquids to selectively extract dyes from waste fibres and transfers the dyes to new fabrics (DyeRecycle, n.d). This allows for a double impact: making textile waste circular, while reducing the environmental impact that occurs during the virgin dyeing process. Additionally, this technology can also convert textile waste scraps into dyestuff powder. The recycling of textiles waste becomes more efficient when allowing these technological innovations to be part of the recycling model. Likewise, Officina39 has developed a technology called Recycrom, an innovative sustainable dyeing solution made from waste textile scraps and uses a process that reduces applications via a chemical solution. The method of application is also a crucial factor here, as suspension application uses pigments which are in fact lightfast and do not breakdown when coming into contact with sunlight. (Mogilireddy, 2018). During this process, scraps of fibre are crystalised into pigmented powders which are re-used as colourants for new products (Recycrom, n.d).

Conclusion

Technology such as IoT, blockchain and communicative systems, are being utilised in other industries such as finance and food, yet the productivity of global supply chain networks for the fashion and textiles sector would benefit from being more interconnected. Technological innovation can facilitate and assist a circular system, notably at the micro level, where the types of technology discussed in this chapter, are a key proponent in managing the life of textiles usage. As the UK and EU policy directives highlight the viability of technical solutions, policy makers also need to gain a more comprehensive understanding of how this can assist circular solutions in the global value chain, where obstacles to adopting activities to enhance circularity, are not entirely obvious to governing bodies. Post-consumer waste recycling also demonstrates barriers, a prominent factor is being able to scale up. The current technology and facilities to accommodate the amounts of textiles in the market, requires financial investments and infrastructure. Second and third life models, such as reuse and repair, also require the ease of product transition, by making these models embedded into the supply chain process, consumers and producers taking on responsibility of managing the life of textile products.

References

Agarwal, P. (2014). *Stock Keeping Unit*. Available at: https://www.fibre2fashion.com/industry-article/7351/stock-keeping-unit#:~:text=Stock%20keeping%20unit%20is%20ordinarily,and%20style%20is%20one%20SKU. (Accessed: 6 August 2022).

Boom Cárcamo, E.A. and Peñabaena-Niebles, R. (2022). Opportunities and challenges for the waste management in emerging and frontier countries through industrial symbiosis. *Journal of Cleaner Production*, 363. doi: https://doi.org/10.1016/j.jclepro.2022.132607

Bennett, E. (2021). IBM newsroom. *Sustainable Supply Chain Optimisation*. Available at: https://uk.newsroom.ibm.com/2021-08-04-Sustainable-Supply-Chain-Optimisation (Accessed: 18th April 2022).

Butner, K. (2010). The smarter supply chain of the future. *Strategy & Leadership*, 38(1): 22-31. Available at: https://www.ibm.com/downloads/cas/AN4AE4QB (Accessed: 20th April 2022).

CB Insights (2022). *Research Briefs: Here are the 5 Biggest Challenges for Brands and Retailers this year – and the tech that can solve them*. Available at: https://www.cbinsights.com/research/consumer-retail-tech-solutions-guide/ (Accessed: 2nd August 2022).

Chen, W. and Lin, C. (2021). Interrelationship among CE adoption obstacles of supply chain in the textile sector: Based on the DEMATEL-ISM Approach. *Mathematics* (Basel), 9(12): 1425. doi: https://doi.org/10.3390/math9121425

Circular ID Initiative (2021). *The Circular Product Data Protocol*. V1.0. Available at: https://www.circulardataprotocol.org/download (Accessed: 3rd May 2022).

Cura, K., Rintala, N., Kamppuri, T., Saarimaki, E. and Heikkila, P. (2021). Textile recognition and sorting for recycling at an automated line using near infrared spectroscopy. *Recycling*, 6(1). doi: http://dx.doi.org/10.3390/recycling6010011.

Demestichas, K. and Daskalakis, E. (2020). Information and communication technology solutions for the circular economy. *Sustainability MDPI* (Basel, Switzerland), 12(18): 7272. doi: https://doi.org/10.3390/su12187272

DyeRecycle, no date. *We give Dyes and Fibres a Second Chance*. Available at: https://www.dyerecycle.com/ (Accessed: 18th April 2022).

Eco-standard Org. (2022). *Make-or-break Aspects of the EU´s Sustainable Products Initiative*. Available at: https://ecostandard.org/news_events/sustainable-products-initiative-a-new-digital-product-passport/ (Accessed: 16th April 2022).

Ertz, M., Sun, S., Boily, E., Kubiat, P. and Quenum, G.G.Y. (2022). How transitioning to Industry 4.0 promotes circular product lifetimes. *Industrial Marketing Management*, 101: 125-140. doi: https://doi.org/10.1016/j.indmarman.2021.11.014.

European Commission (2020). *Circular Economy Action Plan*. Luxembourg: Publications Office. Available at: https://eur-lex.europa.eu/legal-content/EN/TXT/?qid=1583933814386&uri=COM:2020:98:FIN (Accessed: 20th February 2022)

European Commission (2022a). *European Green Deal*. Brussels. Available at: https://ec.europa.eu/commission/presscorner/detail/en/ip_22_2013 (Accessed: 16th April 2022).

European Commission (2022b). *EU Strategy for Sustainable and Circular Textiles*. Brussels. Available at: https://eur-lex.europa.eu/legal-content/EN/TXT/?uri=CELEX%3A52022DC0141 (Accessed: 16th April 2022).

Evreka, no date. *Software Solutions Waste Management for EPR and PRO*. Available at: https://evreka.co/solutions/epr-management/ (Accessed: 30th July 2022).

Greenhouse Gas Protocol (2004). Revised edition. A *Corporate Accounting and Reporting Standards.* Available at: https://ghgprotocol.org/corporate-standard (Accessed: 15th June 2020).

Gray, S. (2022). Interviewed by Sajida Gordon. *Textiles 2030 Data and Policy.* Online United Kingdom.

Järvinen, S., Mäkelä, S., Häikiö, J., Karell, E. (2021). *Clothing Circulator – Data to Extend the Lifetime of Garments.* Verlag der Technischen Universität Graz. Proceedings of the 20th European Roundtable on Sustainable Consumption and Production. doi: https://doi.org/10.3217/978-3-85125-842-4-09

Le, K. (2018). *Textiles Recycling Technologies, Colouring and Finishing Methods.* University of British Columbia. Solid Waste Services, Vancouver. Available at: https://sustain.ubc.ca/sites/default/files/201825%20Textile%20Recycling%20Technologies%2C%20Colouring%20and%20Finishing%20Methods_Le.pdf (Accessed: 4th April 2022).

McKinsey and Company (2020). *Fashion Industry on Climate: How the Fashion Industry can urgently Act to Reduce Its Greenhouse Gas Emissions.* Available at: https://www.mckinsey.com/~/media/mckinsey/industries/retail/our%20insights/fashion%20on%20climate/fashion-on-climate-full-report.pdf (Accessed: 18th May 2022).

Menon, S. (2022). How to use QR codes on food packaging: The all-in-one guide. *Beaconstac,* 11th July. Available at: https://blog.beaconstac.com/2020/12/qr-codes-food-packaging/ (Accessed: 6th August 2022).

Mogilireddy, V. (2018). *Sustainable Dyeing Innovations: Greener Ways to Colour Textiles.* Available at: https://www.prescouter.com/2018/11/sustainable-dyeing-innovations-greener-ways-color-textiles/ (Accessed: 28th June 2022).

Moretto, A. and Macchion, L. (2022). Drivers, barriers, and supply chain variables influencing the adoption of the blockchain to support traceability along fashion supply chains. *Operations Management Research.* doi: https://doi.org/10.1007/s12063-022-00262-y

Pandey, V., Pant, M. and Snasel, V. (2022). Blockchain technology in food supply chains: Review and bibliometric analysis. *Technology in Society,* 69. doi: https://doi.org/10.1016/j.techsoc.2022.101954

Recycrom, no date. *About Recycrom Tech Explained.* Available at: https://recycrom.com/tech-explained/ (Accessed: 4th May 2022).

Red Hat. no date. *Our Company.* Available at: https://www.redhat.com/en/about/company. (Accessed: 19th April 2022).

Sandvik, I.M. and Stubbs, W. (2019). Circular fashion supply chain through textile-to-textile recycling. *Journal of Fashion Marketing and Management,* 23(3): 366-381. doi: https://doi.org/10.1108/JFMM-04-2018-0058

Scheiner, D. (2022). Digital product passport - Anchoring product data on the tangle for a green circular economy. *IOTA Org,* 23rd June. Available at: https://blog.iota.org/digital-product-passport/ (Accessed: 15th July 2022).

Sensoneo, no date. *Smart Waste Monitoring.* Available at: https://sensoneo.com/smart-waste-monitoring/ (Accessed: 28th March 2022).

Smart Fibre Sorting, no date. *Closing the Loop in the Textiles Industry.* Available at: https://smartfibersorting.com/interreg-project-fibersort-closing-the-loop-in-the-textiles-industry-2/ (Accessed: 18th April 2022).

Smith, G. (2021). *IBM Supply Chain Transformation.* IBM Redbooks.

Tossavainen, M. (2019). Efficient textile recycling with NIR Spectroscopy. *Spectral*

Engines, 5th September. Available at: https://www.spectralengines.com/blog/efficient-textile-recycling-with-nir-spectroscopy#:~:text=NIR%20spectroscopy%20is%20based%20on,fibre%20type%20or%20blend%20combination (Accessed: 29th March).

Tripathi, G., Tripathi Nautiyal, V., Ahad, M.A. and Feroz, N. (2021). Blockchain technology and fashion industry – Opportunities and challenges. *Blockchain Technology: Applications and Challenges*, pp. 201-220. Cham: Springer International Publishing.

UK Department for Digital, Culture, Media and Sport, DDCMS. (2022). *UK Digital Strategy*. Available at: https://www.gov.uk/government/publications/uks-digital-strategy/uk-digital-strategy (Accessed: 9th July 2022).

UK Department of Environment, Food and Rural Affairs, DEFRA (2020). *Circular Economy Package*. Available at: https://www.gov.uk/government/publications/circular-economy-package-policy-statement/circular-economy-package-policy-statement#executive-summary (Accessed: 2nd March 2022).

UK Department of Environment, Food and Rural Affairs, DEFRA (2021). *UK Waste Prevention Programme*, WPP. Available at: https://consult.defra.gov.uk/waste-and-recycling/waste-prevention-programme-for-england-2021/supporting_documents/Waste%20Prevention%20Programme%20for%20England%20%20consultation%20document.pdf (Accessed: 23rd February 2022).

UKFT (2021). *Textiles 2030 Roadmap and Circularity Pathway*. Available at: https://www.ukft.org/textiles-2030-roadmap-circularity-pathway/#:~:text=Textiles%202030%20environmental%20targets%20are,by%202050%20at%20the%20latest. (Accessed: 2nd August).

Waldern, J., Stienbrecher, A. and Marinkovic, M. (2021). Digital product passports as enabler of the circular economy. *Chemie Ingenieur Technik: Wiley online library*, 93(11): 1717-1727. doi: https://doi.org/10.1002/cite.202100121

Weiland Textiles, no date. *Fibersort Ready to Start Valorizing Global Textile Waste, Sorting 900 kgs of Post-consumer Textiles per hour, Enabling a Closed Textiles Loop*. Available at: https://www.wieland.nl/en/fibersort-ready-to-start-valorizing-global-textile-waste-sorting-900-kgs-of-post-consumer-textiles-per-hour-enabling-a-closed-textiles-loop/ (Accessed: 18th April 2022).

Williams, A. and Hodges, N. (2022). Signaling sustainability: Exploring consumer perspectives on communicating apparel sustainability information. *Journal of Sustainable Marketing*, 3(1): 26-40. Doi: https://doi.org/10.51300/jsm-2022-49

Wrap, no date. *Resources and Waste strategy for England*. Available at: https://wrap.org.uk/taking-action/collections-recycling/delivering-for-government/resources-waste-strategy-england. (Accessed: 2nd February 2022).

Wrap (2021). *Textile 2030: Circularity Pathway*. Available at: https://wrap.org.uk/sites/default/files/2021-04/Textiles%202030%20Circularity%20Pathway.pdf (Accessed: 28th March 2022).

Zhou, C., Han, G., Via, B.K., Song, Y., Gao, S. and Jiang, W. (2019a). Rapid identification of fibers from different waste fabrics using the near-infrared spectroscopy technique. *Textiles Research Journal*, 89(17): 3610-3616. doi: https://doi.org/10.1177/0040517518817043

Zhou, J., Yu, L., Ding, Q. and Wang, R. (2019b). Textile fibre identification using near-infrared spectroscopy and pattern recognition. *AUTEX Research Journal*, 19(2): 201-209. doi: https://doi.org/10.1515/aut-2018-0055

Index

Contributors Biographies

Dr. Burcikova Mila is a researcher at Centre for Sustainable Fashion, London College of Fashion, UK. At the Centre, she works across a range of research and knowledge exchange projects with focus on micro and small fashion businesses that offer alternatives to the current fashion system. Her research interrogates the connections between fashion and everyday life within the context of climate emergency.

Dr. Cura Kirsti is a postdoctoral researcher at Aalto University, Finland. Her current research focuses on the role of textile fibre identification in circular textile and fashion sector, and transparency and traceability in textile value chains. Dr. Cura is a member of the Fashion/Textile Futures research group at Aalto University.

Gordon Sajida is a lecturer in Fashion Management and researcher at Nottingham Trent University, UK. She is a member of Clothing Sustainability Research Group at NTU. Prior to this her expertise within the fashion industry were at all market levels in product development, garment technology and manufacturing.

Professor in clothing and sustainability, **Grimstad Klepp Ingun**, has been working with clothing at SIFO, Norway since 1999. Her research topics include wearing habits, laundry, lifespan, product development, wool, local clothes, and value chains. Klepp wants to make the knowledge count in the debate on sustainability and actively shares her research through media, exhibitions, debates, and popular books.

Dr. Han Sara is a Senior Lecturer at Nottingham Trent University, UK. As a fashion sustainability professional, currently working in research and education, she has an insightful perspective on effective strategies and tactics to engage stakeholders in meeting clear sustainability targets. Her interests lie in combining skills and knowledge in sustainability, communication, and reporting to provide significant advantages in successfully managing and delivering a strategic approach to enhance the way circular systems are designed to drive positive behaviour change.

Senior Scientist and Project Manager **Heikkilä Pirjo** (D.Sc. Tech) has altogether over 22 years' experience in textile research. Since 2015 she has mainly focused on topics related to textile recycling and circular economy, and in this field she has coordinated several national Finnish jointly funded research projects, including Telaketju research projects. Heikkilä works in VTT Technical Research Center of Finland Ltd.

Horn Susanna DSc (Econ) is a group manager in the Industry and value chains in the Finnish Environment Institute (Syke). Susanna is currently working on projects related to circular economy and life cycle approaches related to textiles, plastics, digitalisation, metals, as well as ecodesign, innovation, policies and data questions. Her main expertise is in the strategic use of life cycle methods. Prior to working in Syke, she has worked in the metals and mining sector in sustainability and innovation positions, as well as in the university sector as a researcher. She has PhD in business-related application of LCA methodologies, Master's degrees in Economics (JYU) and Sustainable Resource Management (TUM).

Jain Sheenam did her PhD at the University of Borås, Sweden in year 2020 as a part of the Erasmus Mundus Joint Doctorate Program, "Sustainable Management and Design for Textile" (SMDTex), funded by the European Commission and the Chinese Scholarship Council (CSC). Her doctoral thesis "Big Data Management Using Artificial Intelligence in the Apparel Supply Chain: Opportunities and Challenges" examines how big data management and artificial intelligence can be used as valuable resources for the apparel supply chain to gain competitive advantage. In years 2021-2022, Sheenam Jain worked as a post doctoral researcher in Aalto University Finland focusing on data aspects in a circular economy.

Laitala Kirsi, PhD, is a Senior Researcher at SIFO, Norway, working with sustainability and textile consumption. Her research themes include clothing lifetimes, quality, maintenance, environmental issues, design, as well as fit and size issues. She uses interdisciplinary research methods based on her educational background in textile engineering (MSc), PhD in Product Design and long experience working with social science research methods.

Løvbak Berg Lisbeth, Research Assistant at SIFO, Norway, works on the WOOLUME, Wasted Textiles, CHANGE and Lasting projects. She is a designer, consultant and fashion futurist with an MA from London College of Fashion, examining the future role of the designer through consumer research and scenario building.

Dr. Mäkelä Mikko is a Research Professor in Intelligent Biomass Processing at VTT Technical Research Center of Finland Ltd. His research includes textile sorting and recycling applications within a circular economy for textiles. He focuses on empirical measurements and data-driven mathematical models using applied spectroscopy, imaging, chemometrics, and machine learning with an interdisciplinary background in environmental and process engineering.

Dr. Niinimäki Kirsi is Associate Professor in Design, especially in Fashion Research in the Department of Design at Aalto University, Finland. Niinimäki approaches the subject of sustainability from a variety of perspectives and her scientific results form a cohesive collection of new knowledge in the sustainable fashion field. Her research group – Fashion/Textile Futures – at Aalto University is involved in several significant research projects that integrate the closed loop, bio-economy and circular economy approaches into fashion and textile systems and expand the understanding of strategic sustainable design.

Peirson-Smith, Anne PhD is Professor of Fashion, School of Design, Northumbria University, UK. She teaches and researches fashion marketing and sustainable fashion management, with an industry background in branding and public relations. She has published numerous articles and book chapters on marketing sustainable fashion and is associate editor of the *Journal of Fashion, Style and Popular Culture* (Intellect Books) and *The Journal of Global Fashion Marketing*. She co-authored *Public Relations in Asia Pacific: Communicating Effectively Across Cultures* (John Wiley, 2010); *Global Fashion Brands: Style, Luxury & History* (Intellect Books, 2014), *Transglobal Fashion Narratives* (Intellect Books, 2018) and *The Fashion Business Reader* (Berg/Fairchild Publishing, 2019).

Ræbild Ulla, is an educated designer and Associate Professor PhD at Design School Kolding, Denmark. Her research lies between sustainability, fashion design practice and education with focus on new roles for designers in companies, organizations, and society for green transition. Ræbild has led the development of Design for Planet MA at DSKD Design School Kolding, Denmark.

Riisberg Vibeke, is Emerita, Associate Professor PhD at Design School Kolding, Denmark. Her research includes design for sustainability with focus on textile and fashion, education and changing roles for designers, design aesthetics, user experience and service systems. Riisberg has been teaching design for sustainability since 1992 and initiated research in design for sustainability at DSKD.

For Product Safety Concerns and Information please contact our EU
representative GPSR@taylorandfrancis.com
Taylor & Francis Verlag GmbH, Kaufingerstraße 24, 80331 München, Germany